住房城乡建设部土建类学科专业"十三五"规划教材
高校建筑学专业规划推荐教材

PROFESSIONAL 建筑学专业英语 （第二版）

王一 岑伟 金倩 编著

ENGLISH IN ARCHITECTURE

中国建筑工业出版社

图书在版编目（CIP）数据

建筑学专业英语/王一，岑伟，金倩编著.—2版.—北京：中国建筑工业出版社，2018.8（2024.1重印）
住房城乡建设部土建类学科专业"十三五"规划教材
高校建筑学专业规划推荐教材
ISBN 978-7-112-22394-7

Ⅰ.①建… Ⅱ.①王…②岑…③金… Ⅲ.①建筑学－英语－高等学校－教材 Ⅳ.①TU

中国版本图书馆CIP数据核字（2018）第138381号

责任编辑：杨 琪 陈 桦
责任校对：李美娜

为了更好地支持相应课程的教学，我们向采用本书作为教材的教师提供课件，有需要者可与出版社联系。
建工书院 http://edu.cabplink.com
邮箱：jckj@cabp.com.cn 电话：(010)58337285
教师QQ群：563612069

住房城乡建设部土建类学科专业"十三五"规划教材
高校建筑学专业规划推荐教材

建筑学专业英语（第二版）

王 一 岑 伟 金 倩 编著

*

中国建筑工业出版社出版、发行（北京海淀三里河路9号）
各地新华书店、建筑书店经销
北京科地亚盟排版公司制版
建工社（河北）印刷有限公司印刷

*

开本：787×1092毫米 1/16 印张：20¼ 字数：502千字
2018年8月第二版 2024年1月第十二次印刷
定价：**49.00**元（赠教师课件）
ISBN 978-7-112-22394-7
（32209）

版权所有 翻印必究
如有印装质量问题，可寄本社退换
（邮政编码100037）

Forwards 前言

《建筑学专业英语》是建筑学专业本科及相关专业学生在大学基础英语学习阶段之后，为进一步学习和提高在本专业领域内以英语为工具进行阅读、表达的综合技能而使用的教材。

本教材的编写以建筑学专业英语在教学内容和教学方式上的创新探索为出发点，力求外语教学内容同建筑学专业培养要求相结合，紧密结合建筑学专业应用的实际需要，对学生进行以阅读和表达为核心的语言训练，促进学生在本专业领域内以英语为媒介获取信息和表达能力的提高。同时结合语言教学拓宽学生的专业知识面，以适应建筑学专业教育的国际化趋势和学生进一步的专业发展。

全书共36个单元，能满足两个学期36周的使用要求。每个单元由四个方面的内容构成：①专业文献精读（Intensive Reading）；②专业文献泛读（Extensive Reading）；③写作、翻译或口语表达技巧（Tips for Writing, Translation and Oral Presentation）；④视听训练（Media Analysis）。视听训练部分由教师根据实际教学情况灵活掌握上课内容，不限于本书推荐的范围。

在教材编写过程中，得到了同济大学建筑与城市规划学院及建筑系的大力支持，教师同行的诚恳建议，以及同济大学建筑学专业和历史建筑保护工程专业学生的积极参与和配合，在此一并表示感谢。

本教材可以供建筑学专业本科及相关专业使用，建筑学专业硕士研究生的专业英语教学亦可选用本教材的部分内容。

由于编著者水平有限，有错误或不当之处欢迎读者不吝赐教。

编著者

Contents 目录

Unit 01　A CITY IS NOT A TREE (PART Ⅰ) ·· 1
　　Section 1　Intensive Reading ·· 1
　　Section 2　Extensive Reading ·· 3
　　Section 3　Tips for Translation ·· 5

Unit 02　A CITY IS NOT A TREE (PART Ⅱ) ·· 7
　　Section 1　Intensive Reading ·· 7
　　Section 2　Extensive Reading ·· 9
　　Section 3　Tips for Translation ·· 11

Unit 03　A CITY IS NOT A TREE (PART Ⅲ) ·· 13
　　Section 1　Intensive Reading ·· 13
　　Section 2　Extensive Reading ·· 15
　　Section 3　Tips for Translation ·· 16

Unit 04　A CITY IS NOT A TREE (PART Ⅳ) ·· 19
　　Section 1　Intensive Reading ·· 19
　　Section 2　Extensive Reading ·· 22
　　Section 3　Tips for Oral Presentation ·· 23

Unit 05　A CITY IS NOT A TREE (PART Ⅴ) ·· 25
　　Section 1　Intensive Reading ·· 25
　　Section 2　Extensive Reading ·· 27
　　Section 3　Tips for Translation ·· 29

Unit 06　A CITY IS NOT A TREE (PART Ⅵ) ·· 31
　　Section 1　Intensive Reading ·· 31
　　Section 2　Extensive Reading ·· 33
　　Section 3　Tips for Translation ·· 34

Unit 07　A CITY IS NOT A TREE (PART Ⅶ) ·· 35
　　Section 1　Intensive Reading ·· 35
　　Section 2　Extensive Reading ·· 38
　　Section 3　Tips for Translation ·· 41

Unit 08　A CITY IS NOT A TREE　(PART VIII) ·········· 43
　　Section 1　Intensive Reading ·········· 43
　　Section 2　Extensive Reading ·········· 45
　　Section 3　Tips for Oral Presentation ·········· 49
　　Section 4　Tips for Writing ·········· 50

Unit 09　A CITY IS NOT A TREE　(PART IX) ·········· 51
　　Section 1　Intensive Reading ·········· 51
　　Section 2　Extensive Reading ·········· 56
　　Section 3　Tips for Translation ·········· 58
　　Section 4　Listening Practice ·········· 59

Unit 10　Modern Housing Prototypes (Part I) ·········· 61
　　Section 1　Intensive Reading ·········· 61
　　Section 2　Extensive Reading ·········· 64
　　Section 3　Tips for Translation ·········· 66

Unit 11　Modern Housing Prototypes (Part II) ·········· 67
　　Section 1　Intensive Reading ·········· 67
　　Section 2　Extensive Reading ·········· 69
　　Section 3　Tips for Translation ·········· 72

Unit 12　Modern Housing Prototypes (Part III) ·········· 75
　　Section 1　Intensive Reading ·········· 75
　　Section 2　Extensive Reading ·········· 78
　　Section 3　Tips for Oral Presentation ·········· 84

Unit 13　Modern Housing Prototypes (Part IV) ·········· 87
　　Section 1　Intensive Reading ·········· 87
　　Section 2　Extensive Reading ·········· 92
　　Section 3　Tips for Translation ·········· 95

Unit 14　Modern Housing Prototypes (Part V) ·········· 97
　　Section 1　Intensive Reading ·········· 97
　　Section 2　Extensive Reading ·········· 101
　　Section 3　Tips for Translation ·········· 103

Unit 15　Modern Housing Prototypes (Part VI) ·········· 105
　　Section 1　Intensive Reading ·········· 105
　　Section 2　Extensive Reading ·········· 110
　　Section 3　Tips for Translation ·········· 112

Unit 16	**Modern Housing Prototypes (Part Ⅶ)**	115
Section 1	Intensive Reading	115
Section 2	Extensive Reading	118
Section 3	Tips for Oral Presentation	119
Unit 17	**Modern Housing Prototypes (Part Ⅷ)**	121
Section 1	Intensive Reading	121
Section 2	Extensive Reading	124
Section 3	Tips for Translation	125
Section 4	Tips for Writing	126
Unit 18	**Modern Housing Prototypes (Part Ⅸ)**	127
Section 1	Intensive Reading	127
Section 2	Extensive Reading	133
Section 3	Tips for Translation	134
Section 4	Listening Practice	135
Unit 19	**Form-Resistant Structures (Part Ⅰ)**	137
Section 1	Intensive Reading	137
Section 2	Extensive Reading	138
Section 3	Tips for Translation	140
Unit 20	**Form-Resistant Structures (Part Ⅱ)**	143
Section 1	Intensive Reading	143
Section 2	Extensive Reading	148
Section 3	Tips for Oral Presentation	149
Unit 21	**Form-Resistant Structures (Part Ⅲ)**	151
Section 1	Intensive Reading	151
Section 2	Extensive Reading	155
Section 3	Tips for Translation	156
Unit 22	**Form-Resistant Structures (Part Ⅳ)**	159
Section 1	Intensive Reading	159
Section 2	Extensive Reading	162
Section 3	Tips for Translation	163
Unit 23	**Form-Resistant Structures (Part Ⅴ)**	165
Section 1	Intensive Reading	165
Section 2	Extensive Reading	169
Section 3	Tips for Translation	172

Unit 24　Form-Resistant Structures （Part Ⅵ） ……………………………… 175
　　Section 1　Intensive Reading ……………………………………………… 175
　　Section 2　Extensive Reading …………………………………………… 178
　　Section 3　Tips for Oral Presentation ……………………………………… 182

Unit 25　Form-Resistant Structures （Part Ⅶ） ……………………………… 183
　　Section 1　Intensive Reading ……………………………………………… 183
　　Section 2　Extensive Reading …………………………………………… 185
　　Section 3　Tips for Translation ……………………………………………… 189

Unit 26　Form-Resistant Structures （Part Ⅷ） ……………………………… 191
　　Section 1　Intensive Reading ……………………………………………… 191
　　Section 2　Extensive Reading …………………………………………… 194
　　Section 3　Tips for Translation ……………………………………………… 200
　　Section 4　Tips for Writing ………………………………………………… 200

Unit 27　Form-Resistant Structures （Part Ⅸ） ……………………………… 203
　　Section 1　Intensive Reading ……………………………………………… 203
　　Section 2　Extensive Reading …………………………………………… 205
　　Section 3　Tips for Translation ……………………………………………… 209
　　Section 4　Listening Practice ……………………………………………… 209

Unit 28　International architectural design competition （Part Ⅰ） …………… 211
　　Section 1　Intensive Reading ……………………………………………… 211
　　Section 2　Extensive Reading …………………………………………… 213
　　Section 3　Tips for Oral Presentation ……………………………………… 214

Unit 29　International architectural design competition （Part Ⅱ） …………… 217
　　Section 1　Intensive Reading ……………………………………………… 217
　　Section 2　Extensive Reading …………………………………………… 219
　　Section 3　Tips for Translation ……………………………………………… 221

Unit 30　International architectural design competition （Part Ⅲ） …………… 223
　　Section 1　Intensive Reading ……………………………………………… 223
　　Section 2　Extensive Reading …………………………………………… 226
　　Section 3　Tips for Translation ……………………………………………… 227

Unit 31　International architectural design competition （Part Ⅳ） …………… 229
　　Section 1　Intensive Reading ……………………………………………… 229
　　Section 2　Extensive Reading …………………………………………… 231
　　Section 3　Tips for Translation ……………………………………………… 233

Unit 32　International architectural design competition（Part Ⅴ） ………… 235
　　Section 1　Intensive Reading ……………………………………………… 235
　　Section 2　Extensive Reading ……………………………………………… 238
　　Section 3　Tips for Oral Presentation …………………………………… 239

Unit 33　"Im Birch" School（Part Ⅰ） ……………………………………… 241
　　Section 1　Intensive Reading ……………………………………………… 241
　　Section 2　Extensive Reading ……………………………………………… 250
　　Section 3　Tips for Translation …………………………………………… 254

Unit 34　"Im Birch" School（Part Ⅱ） ……………………………………… 257
　　Section 1　Intensive Reading ……………………………………………… 257
　　Section 2　Extensive Reading ……………………………………………… 263
　　Section 3　Tips for Translation …………………………………………… 267

Unit 35　Detached Family Home（Part Ⅰ） ………………………………… 269
　　Section 1　Intensive Reading ……………………………………………… 269
　　Section 2　Extensive Reading ……………………………………………… 274
　　Section 3　Tips for Translation …………………………………………… 276
　　Section 4　Tips for Writing ………………………………………………… 277

Unit 36　Detached Family Home（Part Ⅱ） ………………………………… 279
　　Section 1　Intensive Reading ……………………………………………… 279
　　Section 2　Extensive Reading ……………………………………………… 283
　　Section 3　Tips for Oral Presentation …………………………………… 286
　　Section 4　Listening Practice ……………………………………………… 286

Appendix ……………………………………………………………………………… 289
　　Vocabulary 词汇 …………………………………………………………… 290
　　Notes for Video Material 音像资料注解 ………………………………… 306
　　Bibliography 参考文献 …………………………………………………… 315

Unit 01

Section 1
Intensive Reading

A City Is Not a Tree
Christopher Alexander

Part I

Introduction Many design professionals admire and attempt to incorporate into their designs for the built environment elements reflecting underlying human psychological and spiritual needs and cultural values. But none has broken so completely with conventional architectural practice and sought more deeply to make his designs reflect these fundamental values than Austrian-born, British-trained, U. S. -based architect/planner Christopher Alexander.

Alexander is a self-proclaimed iconoclast, deliberately distancing himself from virtually all the major mainstream currents of twentieth-century architectural and planning thought. It is notable that the eight "treelike" plans he singles out for attack in the following selection represent a diverse set of the most respected and famous twentieth-century plans from Le Corbusier's plan for the new town of Chandigarh, India, based on his principles for a contemporary city, to Paolo Soleri's visionary megastructure of Mesa City in the Arizona desert.

Since publication of his provocative early attack on the sterility of formal "treelike" city plans in the following selection, Alexander has been engaged in a lifelong search to decipher the deep structures underlying human needs and to define recurring patterns for a new paradigm of architecture. The following selection is clear that a city should not be designed with a neatly branching treelike organization dividing functions from each other. Alexander condemns tidy city plans which lay out discretely bounded neighborhoods, zone one area for housing and another for business, or establish areas just for universities or cultural facilities. He sees human activity as much more complex and overlapping than that.

Alexander's approach to describing how cities should be designed in this selection may trouble readers who seek clear, rational guidelines. He takes the position that not enough is yet known about how to design non-treelike cities to provide definite answers. Like an artist or a Zen master instructing an apprentice, Alexander closes this selection with provocative analogies, examples, and metaphors. He suggests how an individual might pursue the quest for good design, but he does not offer a stock set of the answers.

During the past three decades Alexander and his colleagues and students at the Center for Environmental Structure at the University of California of Berkeley have conducted a series of "experiments" working to understand and demonstrate how to design cities which are not "trees". Alexander's writings since "A City Is Not a Tree" provide an abundance of specific principles and examples as well as many more unanswered questions and lines for exploration.

While Alexander is fascinated with physical form, his approach begins with an interactive process working with clients to understand their most fundamental needs. Profoundly respectful of the ideas of clients, Alexander's projects incorporate rammed earth and chicken wire into housing for Mexicali slum dwellers and Zen architectural details into a Japanese school. He and his followers seek architecture which is "alive"; architecture that possesses "the quality without a name".

Consider the relevance of J. B. Jackson's description of how the informal vernacular architecture of small U. S. towns meets human needs to Alexander's conviction that built environments that grow organically contain important lessons for planners. Alexander shares architectural critic Jane Jacobs's love of apparently chaotic, jumbled urban neighborhoods. Like Jacobs he sees a complex order and rationality behind an apparently disorderly facade. Consider Alexander's concept of a semi-lattice structure in relation to Jacobs's argument for designing streets to provide play space for children, security, and areas for human interaction as well as space for cars to drive. A casual observer might consider the resulting street a confused and disorderly one. She might not see how it meets multiple, complex human needs. Alexander would like to help architects and planners design streets which achieve the positive qualities of lively streets in New York's Greenwich Village or Boston's West End before urban renewal tidied up (and deadened) the streetscape. Note also the similarity to British architect/planner Raymond Unwin's respect for natural cities and for urban forms shaped by the ideas of their residents.

Alexander's theories are developed in a series of books published by Oxford University Press in New York: The Oregon Experiment (1975), A Pattern Language (1977), The Timeless Way of

Building (1979), The Linz Cafe (1981), The Production of Houses (1985), and A New Theory of Urban Design (1987). An overview of his work by Ingrid F. King is "Christopher Alexander and Contemporary Architecture" in Architecture and Urbanism (August 1993).

Words and Phrases

1. incorporate into   结合/incorporate with
2. psychological   *adj.*   心理（上）的/spiritual 精神上的
3. built environment   建成环境
4. break with   断交，决裂
5. self-proclaimed   *adj.*   自称的
6. iconoclast   *n.*   反偶像者，提倡打破旧习的人
7. single out   挑选（出）
8. megastructure   *n.*   巨型结构
9. sterility   *n.*   贫乏
10. decipher   *v.*   解码，破解/decode
11. paradigm   *n.*   范例，范型
12. discretely   *adv.*   分离的
13. zone   *n.*   分区
14. overlap   *v.*   重叠，交迭/superimpose
15. Zen   禅（佛教）
16. metaphor   *n.*   隐喻
17. rammed earth   夯土
18. vernacular   *adj.*   本土的，本国的，地方的
19. renewal   *n.*   更新/renovation/revitalization

Section 2
Extensive Reading

Paolo Soleri

Born in Turin, Italy on June 21, 1919, Paolo Soleri was awarded his Ph. D. with highest honors in architecture from the Torino Polytechnico in 1946. He came to the United States in 1947 and spent a-year-and-a-half in fellowship with Frank Lloyd Wright at Taliesin West in Arizona, and at Taliesin East in Wisconsin. During this time, he gained international recognition for a bridge design displayed at the Museum of Modern Art and published in The Architecture of Bridges by Elizabeth Mock.

He returned to Italy in 1950 where he was commissioned to build a large ceramics factory, "Ceramica Artistica Solimene". The processes he became familiar with in the ceramics industry led to

Figure 1-1   Paolo Soleri

his award-winning designs of ceramic and bronze wind bells and silt cast architectural structures. For over 30 years, the proceeds from the wind bells have provided funds for construction to test his theoretical work (Figure 1-1).

In 1956 he settled in Scottsdale, Arizona, with his wife, Colly, and their two daughters. Dr. and Mrs. Soleri made a life-long commitment to research and experimentation in urban planning, establishing the Cosanti Foundation, a not-for-profit educational foundation.

The Foundation's major project is Arcosanti, a prototype town for 5,000 people designed by Soleri, under construction since 1970. Located at Cordes Junction, in central Arizona, the project is based on Soleri's concept of "Arcology," architecture coherent with ecology. Arcology advocates cities designed to maximize the interaction and accessibility associated with an urban environment; minimize the use of energy, raw materials and land, reducing waste and environmental pollution; and allow interaction with the surrounding natural environment.

A landmark exhibition, "The Architectural Visions of Paolo Soleri," organized in 1970 by the Corcoran Gallery of Art in Washington, DC, traveled extensively in the U.S. and Canada, breaking records for attendance. "Two Suns Arcology, A Concept for Future Cities" opened at the Xerox Square Center in Rochester, New York, in 1976. In 1989 "Paolo Soleri Habitats: Ecologic Minutiae", and exhibition of arcologies, space habitats and bridges, was presented at the New York Academy of Sciences. Most recently, "Soleri's Cities, Architecture for the Planet Earth and Beyond" was featured at the Scottsdale Center for the Arts in Scottsdale, AZ. His work has been exhibited worldwide.

Soleri has received one fellowship from the Graham Foundation and two from the Guggenheim Foundation. He has been awarded three honorary doctorates, the American Institute of Architects Gold Medal for Craftmanship in 1963, the Gold Medal from the World Biennieal of Architecture in Sofia, Bulgaria, in 1981, and the Silver Medal of the Academied Architecture in Paris, 1984. Soleri is a distinguished lecturer in the College of Architecture at Arizona State University.

He has written six books and numerous essays and monographs. When he is not traveling on the international lecture circuit, Soleri divides his time between Cosanti, the original site for his research located in Scottsdale, and Arcosanti.

Resources for Reference

http://www.patternlanguage.com

http://www.arcosanti.org

Section 3
Tips for Translation

General Principles for Translation 翻译的一般原则

19世纪末翻译家严复提出的"信、达、雅",直到今天还常常被作为评价翻译工作是否到位的一般标准。

所谓"信"指的是翻译要忠实于原文。无论是英译汉还是汉译英,其最终目的是让读者了解原文的意思。特别是对于专业英语的翻译,更是必须符合原意,不能有任何篡改。因此合格的翻译必须建立在充分理解原文的基础上,不但要以读懂原文的字面意思为基础,更要力求品味出原文内在的语气、倾向。

应该注意的是,追求翻译中的"信",并不是不知变通的"硬译"、"死译",要避免一味固守词对词、短语对短语、句子对句子、结构对结构的"对译",而丝毫不考虑英汉两种语言在语言习惯和语法特征等方面的固有差异。特别是在汉译英的时候,最不好的习惯是在汉英词典中搜寻汉语词汇对应的英语词汇,并不加分析地直接套用。殊不知同样的汉语词汇由于上下文的不同,真正的含义往往大相径庭,而同样的英语词汇其汉译却常常是不一样的。例如同样一个"保护",在"环境保护"和"历史保护"中的英译就完全不同,同样一个"ambition",可以翻译成"野心",也可以翻译成"志向",这样的例子也是不一而足。

所谓"达"通常是指翻译要通顺。英译汉和汉译英要达到"达"的要求,必须符合译文的语言习惯。例如,在英译汉时要体现"达",就必须把英语原文翻译成合乎汉语规范和习惯的文字,也就是说译文必须是明白通畅的现代汉语。例如,有人将"His addition completed the list"翻译成"他的加入结束了名单"或"他的加入完成了名单",十分别扭,根本不符合汉语的表达习惯。关于上句,如果翻译为"把他添上,名单就全了"不但容易看懂,读起来也顺口多了。

"信"、"达"、"雅"三个标准,是从易到难的。可以说"雅"就是翻译的最高要求和最高境界。所谓"雅",就是要使译文流畅,有文采。要达到"雅"的标准,译者必须同时具备很高的英语和汉语修养。从建筑学专业英语的角度来看,翻译的最基本要求是"信"和"达",即力求准确、流畅。

Unit 02

Section 1
Intensive Reading

A City Is Not a Tree
Christopher Alexander

Part II

A City Is Not a Tree

The tree of my title is not a green tree with leaves. It is the name for a pattern of thought. The semi-lattice is the name for another, more complex, pattern of thought.

In order to relate these abstract patterns to the nature of the city, I must first make a simple distinction. I want to call those cities which have arisen more or less spontaneously over many, many years natural cities. And I shall call those cities and parts of cities which have been deliberately created by designers and planners artificial cities. Siena, Liverpool, Kyoto, Manhattan are examples of natural cities. Levittown, Chandigarh, and the British New Towns are examples of artificial cities.

It is more and more widely recognized today that there is some essential ingredient missing from artificial cities. When compared with ancient cities that have acquired the patina of life, our modern attempts to create cities artificial are, from a human point of view, entirely unsuccessful.

Architects themselves admit more and more freely that they really like living in old buildings more than new ones. The non-art-loving public at large, instead of being grateful to architects for what they do, regards the onset of modern buildings and modern cities everywhere as an inevitable, rather sad piece of the larger fact that the world is going to the dogs.

It is much too easy to say that these opinions represent only people's unwillingness to forget the past, and their determination to be traditional. For myself, I trust this conservatism. Americans are usually willing to move with the times. Their growing reluctance to accept the modern city evidently expresses a longing for some real thing, something which for the moment escapes our grasp.

The prospect that we may be turning the world into a place peopled only by little glass and concrete boxes has alarmed many architects too. To combat the glass box future, many valiant protests and designs have been put forward, all hoping to recreate in modern form the various characteristics of the natural city which seem to give it life. But so far these designs have only remade the old. They have not been able to create the new.

"Outrage", the Architectural Review's campaign against the way in which new construction and telegraph poles are wrecking the English town, based its remedies, essentially, on the idea that the spatial sequence of buildings and open spaces must be controlled if scale is to be preserved——an idea that really derives from Camillo Sitte's book about ancient squares and piazzas.

Another kind of remedy, in protest against the monotony of Levittown, tries to recapture the richness of shape found in the houses of a natural old town. Llewelyn Davies's village at Rushbrooke in England is an example——each cottage is slightly different from its neighbor, the roofs out in and out at picturesque angles.

A third suggested remedy is to get high density back into the city. The idea seems to be that if the whole metropolis could only be like Grand Central Station, with lots and lots of layers and tunnels all over the place, and enough people milling around in them, maybe it would be human again.

Words and Phrases

1. lattice   *n.*   网格
2. make a distinction   区别、区分
3. spontaneously   *adv.*   自然地、自发地
4. natural city   自然城市/artificial city   人造城市
5. ingredient   *n.*   成分、要素
6. patina   *n.*   光泽
7. at large   普遍地
8. onset   *n.*   肇始
9. go to the dogs   堕落、潦倒

10. conservatism   *n.*   保守主义
11. reluctance   *n.*   勉强、不愿意
12. valiant   *adj.*   勇敢的
13. Outrage   *n.*   暴行、侮辱
14. wreck   *v.*   破坏、拆毁
15. remedy   *n.*   疗法、解决问题的手段
16. spatial sequence   空间序列

Section 2
Extensive Reading

Passage 1
Levittown

Levittown has long represented the paradigmatic postwar American suburb. Yet very little in the way of good critical work has been done on the history and significance of this American cultural icon. Over the past decade I have been assembling materials to provide an ongoing cultural history of Levittown and, through its story, to offer a more nuanced and sympathetic picture of American suburban life in the Cold War era. Part of a larger project, Outside the Gates: Cultural Landscapes from the Material to the Virtual, my Levittown work has become so interesting in itself that I have allowed it to evolve into something closer to a work of collaborative history, here on the net.

Passage 2
Grand Central Terminal

Grand Central Terminal is one of the two monumental gateways that were built in the heyday of railway transportation.

The monumental railway station was constructed in 1903~1913 for the New York and Harlem Railroad company. It is a grand Beaux-Arts building which serves as a transportation hub connecting train, metro, car and pedestrian traffic in an efficient way. It has 67 train tracks on two different levels.

The other, even grander railway station——the Penn Station——was built in 1902~1911 after a design by Charles McKim. In an act of vandalism, the monumental landmark was destroyed in 1963~1966 and replaced by a banal railway station and office tower. The Grand Central Terminal almost suffered a similar fate but thanks to New York City's new landmark preservation laws, the building was able to escape the wrecking ball.

The current Grand Central Terminal was not the first railway station at 42nd street and Park

Avenue. As early as in 1863 Cornelius Vanderbilt, known as "the Commodore" consolidated railroad lines including the Harlem Railroad and New York Central Railroad. By the end of the decade the need for a large railway station became apparent.

In 1869, Vanderbilt commissioned architect John B. Snook to build the largest railway station in the world on a large property at 42nd street. The Grand Central Station featured a large glass and steel train shed (650ft long, 100ft high and 200ft wide). But increasing traffic and the smoke from the steam engines obscured vision in the Park Avenue tunnel, causing an accident in 1902. 17 People were killed and a public outcry called for electrification of the railway system. This resulted in a new state law requiring that steam engines would not be allowed in Manhattan, starting in 1910.

Shortly after the accident, the New York Central Railroad proposed plans for a larger Grand Central station. The costly electrification and construction of the new railway station was compensated by the use of air rights: Electrification made it possible for the tracks to be paved over all the way to 49th street. Developers were allowed to construct buildings on top of it, but had to pay an extra sum to the railway company, the so-called air rights.

In 1903 a competition was held for the design of the new Grand Central. The firm of Reed and Stem was chosen. William K. Vanderbilt II, one of the descendants of the 'Commodore' asked Warren and Wetmore to collaborate with Reed and Stem. While the latter were responsible for the overall design, Warren and Wetmore were responsible for the architectural details and Beaux-Arts style.

The project included not just the new railway station, but a whole complex with office buildings and apartments, which became known as "Terminal City". This was a "city in the city" complex, similar to the concept of the Rockefeller Center created several decades later. Special attention was paid to the circulation of traffic. Pedestrians and cars are separated by special elevated ramps which lead the cars around the railway station.

Construction of the Grand Central Terminal lasted 10 years and cost 80 million dollars. In the process, 180 buildings between 42nd and 50th street, including hospitals and churches, were demolished. The railway station officially opened on Sunday February 2, 1913. But it would last until 1927 before the station was fully operational.

The building's facade on 42nd Street has a true beaux-arts design. Large arches flanked by Corinthian columns are topped by a large sculpture group designed by Jules-Alexis Coutain. The 50ft high group depicts Mercury (the god of commerce) supported by Minerva and Hercules (representing mental and moral strength). Inside, the main concourse is most impressive. It is 470ft long, 160ft wide and 150ft high. The ceiling is painted by the French artist Paul Helleu. The design with zodiac constellations was taken from a medieval manuscript. Light enters the main concourse through three 75ft arched windows. The western double staircase in Botticino marble was designed after the large staircase in the former Opera building in Paris. It connects the main concourse with the entrance on the Vanderbilt Avenue. The floor of the concourse if of Tennessee marble, the walls of Caen stone.

In 1994, the firms of LaSalle Partners and Williams Jackson Ewing were chosen by the Metropolitan Transportation Authority to redevelop the Grand Central Terminal. The firms were chosen for their successful renovation of another Beaux-Arts icon, the Union Station in Washington DC. The MTA's goal was to increase revenue while restoring the building's former grandeur. This was achieved by renovating the large public areas, removing former alterations (like lowered ceilings), adding a new entrance and creating a retail mall and food court, similar to the renovation project in Washington D. C. During the 197 million dollar restoration process, a large iron eagle was added on top of the new Lexington Avenue & 43rd Street entrance. This eagle once adorned the first Grand Central station in 1898.

Resources for Reference

A Research by Peter Bacon Hales
Art History Department, University of Illinois at Chicago
http://www. uic. edu/ ~ pbhales/Levittown. html
http://www. aviewoncities. com/nyc/grandcentralterminal. htm

Section 3
Tips for Translation

Omission 省略

主要是指代词（人称代词、物主代词）、连接词、冠词和介词等的翻译，不必追求词和词一一对译，从而使翻译更符合汉语的表达习惯。

1. Many design professionals admire and attempt to incorporate into their designs for the built environment elements reflecting underlying human psychological and spiritual needs and cultural values.
不少设计师十分推崇并尝试在设计中结合那些能够反映根本的人类心理、精神需求和文化价值观的建成环境要素。
  上句中的物主代词"their"不必译出。

2. It is notable that the eight "treelike" plans he singles out for attack in the following selection

represent a diverse set of the most respected *and* famous twentieth-century plans from…
值得注意的是：他挑出来在下文中进行攻击的 8 个 "树状" 规划是 20 世纪最受尊重、最负盛名的规划方案的代表，包括……

  上句中的 "and" 没有必要译出。

3. Like *an* artist or *a* Zen master instructing an apprentice, Alexander closes this selection with provocative analogies, examples, and metaphors.
如同指导学徒的艺术家或禅师一样，亚历山大用具有启发性的类比、举例和隐喻来结束这段节选。

  上句中的 "an" 和 "a" 可省略不译。

4. Paolo Soleri was awarded his Ph. D. with highest honors in architecture from the Torino Polytechnico *in* 1946.
1946 年保罗·索拉里以最高荣誉被都灵理工学院授予建筑学博士学位。

  有些表示时间的介词可以不翻译。

5. Smoking is prohibited *in* public space.
公共场合禁止吸烟。

  有些表示地点的介词也可以不翻译。

6. Each season the Metropolitan stages more than two hundred performances of opera in New York. More than 800,000 people attend the performances in the opera house during the season.
每个演出季大都会剧院的演出都超过 200 场，观众人数超过八十万人。

  上句中两个 "season" 翻译一次即可。

Unit 03

Section 1
Intensive Reading

A City Is Not a Tree
Christopher Alexander

Part III

Another very brilliant critic of the deadness which is everywhere is Jane Jacobs. Her criticisms are excellent. But when you read her concrete proposals for what we should do instead, you get the idea that she wants the great modern city to be a sort of mixture between Greenwich Village and some Italian hill town, full of short blocks and people sitting in the street.

The problem the designers have tried to face is real. It is vital that we discover the property of old towns which gave them life and get it back into our own artificial cities. But we cannot do this merely by remaking English villages, Italian piazzas, and Grand Central Stations. Too many designers today seem to be yearning for the physical and plastic characteristics of the past, instead of searching for the abstract ordering principle which the towns of the past happened to have, and which our modern conceptions of the city have not yet found.

What is the inner nature, the ordering principle, which distinguishes the artificial city from the natural city?

You will have guessed from my title what I believe this ordering principle to be. I believe that a natural city has the organization of a semi-lattice; but that when we organize a city artificially, we organize it as a tree.

Both the tree and the semi-lattice are ways of thinking about how a large collection of many small systems goes to make up a large and complex system. More generally, they are both names for structures of sets.

In order to define such structures, let me first define the concept of a set. A set is a collection of elements which for some reason we think of as belonging together. Since, as designers, we are concerned with the physical living city and its physical backbone, we most naturally restrict ourselves to considering sets which are collections of material elements such as people, blades of grass, cars, bricks, molecules, houses, gardens, water pipes, the water molecules that run in them, etc.

When the elements of a set belong together because they cooperate or work together somehow, we call the set of elements a system.

For example, in Berkeley at the corner of Hearst and Euclid, there is a drugstore, and outside the drugstore a traffic light. In the entrance to the drugstore there is a newsrack where the day's papers are displayed. When the light is red, people who are waiting to cross the street stand idly by the light; and since they have nothing to do, they look at the papers displayed on the newsrack which they can see from where they stand. Some of them just read the headlines, others actually buy a paper while they wait.

This effect makes the newsrack and the traffic light interdependent; the newsrack, the newspapers on it, the money going from people's pockets to the dime slot, the people who stop at the light and read papers, the traffic light, the electric impulses which make the lights change, and the sidewalk which the people stand on form a system——they all work together.

From the designer's point of view, the physically unchanging part of this system is of special interest. The newsrack, the traffic light, and the sidewalk between them, related as they are, form the fixed part of the system. It is the unchanging receptacle in which the changing parts of the system-people, newspapers, money, and electrical impulses-can work together. I define this fixed part as a unit of the city. It derives its coherence as a unit both from the forces which hold its own elements together, and from the dynamic coherence of the larger living system which includes it as a fixed invariant part.

Of the many, many fixed concrete subsets of the city which are the receptacles for its systems, and can therefore be thought of as significant physical units, we usually single out a few for special consideration. In fact, I claim that whatever picture of the city someone has is defined precisely by the subsets he sees as units.

Now, a collection of subsets which goes to make up such a picture is not merely an amorphous collection. Automatically, merely because relationships are established among the subsets once the subsets are chosen, the collection has a definite structure.

Words and Phrases

1. square/piazza   *n.*   广场/plaza
2. monotony   *n.*   单调、千篇一律
3. cottage   *n.*   村舍
4. picturesque   *adj.*   独特的、风景如画的
5. metropolis   *n.*   大都市
6. mill   *v.*   转悠
7. yearn for 渴望
8. plastic   *adj.*   造型的
9. distinguish from   区别
10. backbone   *n.*   支柱、脊椎
11. molecule   *n.*   分子
12. newsrack   *n.*   报栏
13. dime slot 投币口/dime 一角硬币
14. receptacle   *n.*   容器
15. dynamic   *adj.*   动态的
16. amorphous   *adj.*   形态不定的

Section 2
Extensive Reading

Camillo Sitte

Camillo Sitte (1843 ~ 1903) is best known among urban planners and architects for his book ***City Planning according to Its Artistic Principles*** from 1889.

Sitte's treatise on the basic questions of urban planning was surprisingly successful. The notion of urban planning as being an essentially public and spatial venture was received euphorically and entered numerous planning ordinances throughout Europe. The 1920s avant-garde, on the other hand, emphatically rejected Sitte's theory. The reception continued to be ambivalent until the 1970s, when architects and urban planners rediscovered the importance of the Viennese theoretician. This development culminated in the "new urbanism" movement which counts Sitte's magnum opus as one of its primary historical references. In the field of historiography, only more recent scholarship was able to award Sitte his befitting position as the leading theoretician of urban design in the late 19th century.

Almost all scientific works on Sitte until today focused exclusively on his theory of urban planning, neglecting a systematic analysis of his voluminous estate, which consists of 60 writings on architecture and urban planning, 60 writings on music, painting, art history and arts and crafts, 19 writings on pedagogy, numerous letters, 25 architectural and 17 urban design projects. Access to Sitte's extensive work in these fields is, however, crucial for any historical and source critical interpretation of his main work.

Resources for Reference

http://www.library.cornell.edu/Reps/DOCS/sitte.htm

Section 3
Tips for Translation

Conversion 词类转译

　　汉语与英语的表达方式和语言习惯在很多情况下有所不同。因此在英语翻译成汉语的过程中，不必拘泥于英文原文某些词的词类，可以通过动词—名词、形容词—副词等的相互转译，力求汉译畅达。

1. I believe that a natural city has the organization of a semi-lattice; but that when we *organize* a city artificially, we *organize* it as a tree.
我相信自然城市的组织方式是半网络状的；而当我们人为地组织城市的时候，往往采取的是一种树状的组织方式。
　　上句中有两个"organize"，为了避免重复，可把后面一个"organize"译成汉语中的名词"组织方式"。

2. Another kind of remedy, in *protest* against the monotony of Levittown, tries to recapture the richness of shape found in the houses of a natural old town.
另外一种解决方案反对利维市的单调乏味，并试图重拾自然的老城中住宅的丰富形式。
　　句中的"protest"（名词）可按动词翻译。

3. The hotel can *room* 150 guests.
该宾馆可以提供150人住宿。
　　从动词（room）转译为名词。

4. There has been dramatic *transformation* of the urban-scape of Shanghai in the last two decades.

最近 20 年来上海城市景观发生了巨大变化。

  从名词（transformation）转译为动词。

5. Different architects addressed the similar issues of historical preservation *differently*.
对于历史保护的类似问题，不同的建筑师采取不同的回答。

  从副词（differently）转译为形容词。

6. I am *sure* they will win the design competition.
我相信他们会赢得设计竞赛。

  从形容词（sure）转译为动词。

Unit 04

Section 1
Intensive Reading

A City Is Not a Tree
Christopher Alexander

Part IV

To understand this structure, let us think abstractly for a moment, using numbers as symbols. Instead of talking about the real sets of millions of real particles which occur in the city, let us consider a simpler structure made of just half a dozen elements. Label these elements 1, 2, 3, 4, 5, 6. Not including the full set [1, 2, 3, 4, 5, 6], the empty set [-], and the one-element sets [1], [2], [3], [4], [5], [6], there are 56 different subsets we can pick from six elements.

Suppose we now pick out certain of these 56 sets (just as we pick out certain sets and call them units when we form our picture of the city). Let us say, for example, that we pick the following subsets: [123], [34], [451], [234], [345], [12345], [3456].

What are the possible relationships among these sets? Some sets will be entirely part of larger sets, as [34] is part of [345] and [3456]. Some of the sets will overlap, like [123] and [234]. Some of the sets will be disjoint——that is, contain no elements in common, like [123] and [45].

We can see these relationships displayed in two ways. In diagram (*a*) (Figure 4 – 1) each set chosen to be a unit has a line drawn round it. In diagram (*b*) the chosen sets are arranged in order of ascending magnitude, so that whenever one set contains another (as [345] contains

[34]), there is a vertical path leading from one to the other. For the sake of clarity and visual economy, it is usual to draw lines only between sets which have no further sets and lines between them; thus the line between [34] and [345], and the line between [345] and [3456], make it unnecessary to draw a line between [34] and [3456].

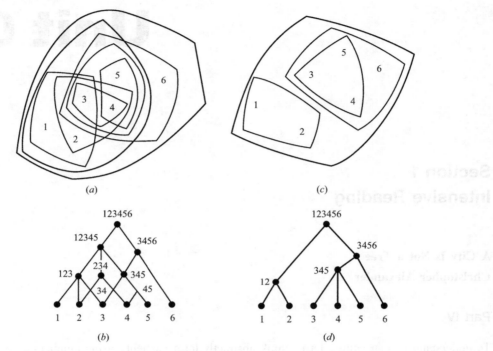

Figure 4-1

As we see from these two representations, the choice of subsets alone endows the collection of subsets as a whole with an overall structure. This is the structure which we are concerned with here. When the structure meets certain conditions it is called a semi-lattice. When it meets other more restrictive conditions, it is called a tree.

The semi-lattice axiom goes like this: A collection of sets forms a semi-lattice if and only if, when two overlapping sets belong to the collection, then the set of elements common to both also belongs to the collection.

The structure illustrated in diagrams A and B is a semi-lattice. It satisfies the axiom since, for instance, [234] and [345] both belong to the collection, and their common part, [34], also belongs to it. (As far as the city is concerned, this axiom states merely that wherever two units overlap, the area of overlap is itself a recognizable entity and hence a unit also. In the case of the drugstore example, one unit consists of the newsrack, sidewalk, and traffic light. Another unit consists of the drugstore itself, with its entry and the newsrack. The two units overlap in the newsrack. Clearly this area of overlap is itself a recognizable unit, and so satisfies the axiom above which defines the characteristics of a semi-lattice.)

The tree axiom states: **A collection of sets forms a tree if and only if, for any two sets that belong to the collection, either one is wholly contained in the ether, or else they are wholly disjoint.**

The structure illustrated in diagrams (c) and (d) is a tree. Since this axiom excludes the possibility of overlapping sets, there is no way in which the semi-lattice axiom can be violated, so that every tree is trivially simple semi-lattice.

However, in this paper we are not so much concerned with die fact that a tree happens to be a semi-lattice, but with the difference between trees and those more genera semi-lattice which are not trees because they do contain overlapping units. We are concerned with the difference between structures in which no overlap occurs, and those structures in which overlap does occur.

It is not merely the overlap which makes the distinction between the two important. Still more important is the fact that the semi-lattice is potentially a much more complex and subtle structure than a tree. We may see just how much more complex semi-lattice can be than a tree in the following fact: a tree based on 20 elements can contain at most 19 further subsets of the 20, while a semi-lattice based on the same 20 elements can contain more than 3,000,000 different subsets.

This enormously greater variety is an index of the great structural complexity a semi-lattice can have when compared with the structural simplicity of a tree. It is this lack of structural complexity, characteristic of trees, which is crippling our conceptions of the city.

To demonstrate, let us look at some modern conceptions of the city, each of which I shall show to be essentially a tree. It will perhaps be useful, while we look at these plans, to have a little ditty in our minds:

Big fleas havelittle fleas
Upon their back to bite'em
Little fleas have lesser fleas
And so ad infinitum

This rhyme expresses perfectly and succinctly the structural principle of the tree.

Words and Phrases

1. ascending magnitude   递增
2. clarity   *n.*   清晰
3. economy   *n.*   简洁

4. endows with 赋予
5. illustrate v. 用插图说明
6. diagram n. 图解、图表
7. axiom n. 公理
8. entity n. 实体
9. subtle adj. 微妙的
10. complexity/simplicity 简单/复杂
11. cripple v. 削弱
12. demonstrate v. 证明
13. ditty n. 小曲、小调
14. rhyme n. 韵律、押韵的诗词
15. succinctly adv. 简洁地

Section 2
Extensive Reading

City Planning According to Artistic Principles（Introduction）

The publication at Vienna in May 1889 of *Der Städtebau nach seinen künstlerischen Grundsätzen* (City Planning According to Artistic Principles) began a new era in Germanic city planning. Its author, Camillo Sitte (1843~1903) strongly criticized the current emphasis on broad, straight boulevards, public squares arranged primarily for the convenience of traffic, and efforts to strip major public or religious landmarks of adjoining smaller structures regarded as encumbering such monuments of the past.

Sitte proposed instead to follow what he believed to be the design objectives of those whose streets and buildings shaped medieval cities. He advocated curving or irregular street alignments to provide ever-changing vistas. He called for T-intersections to reduce the number of possible conflicts among streams of moving traffic. He pointed out the advantages of what came to be know as "turbine squares"——civic spaces served by streets entering in such a way as to resemble a pin-wheel in plan.

His teachings became widely accepted in Austria, Germany, and Scandinavia, and in less than a decade his style of urban design came to be accepted as the norm in those countries. There were, of course, critics of this approach, and ultimately the kind of carefully studied informality that Sitte endorsed came itself to be regarded as old-fashioned. Nevertheless, from 1890 until the outbreak of the First World War, the majority of the numerous extension plans for enlarging the rapidly growing cities of Germany incorporated all or some of the elements so strongly supported by Sitte and his followers.

Resources for Reference

http://www.library.cornell.edu/Reps/DOCS/sitte.htm

Section 3
Tips for Oral Presentation

Oral Communication Is Different from Written Communication

Listeners have one chance to hear your talk and can't "re-read" when they get confused. In many situations, they have heard or will hear several talks on the same day. Being clear is particularly important if the audience can't ask questions during the talk. There are two well-know ways to communicate your points effectively. The first is to K. I. S. S. (keep it simple stupid). Focus on getting one to three key points across. Think about how much you remember from a talk last week. A presentation can easily be ruined if the content is too difficult for the audience to follow or if the structure is too complicated. Give your presentation a simple and logical structure. Include an introduction in which you outline the points you intend to cover and a conclusion in which you go over the main points of your talk. Second, repeat key insights: tell them what you're going to tell them (Forecast), tell them, and tell them what you told them (Summary).

Resources for Reference

http://pages.cs.wisc.edu/~markhill/conference-talk.html

Resources for Reference

http://www.library.cornell.edu/Research/OK-guide.html

Section 3
Tips for Oral Presentation

Oral Communication Is Different from Written Communication

Listeners have one chance to hear your talk and can't "re-read" when they get confused. In most situations, they have heard or will hear several talks on the same day. Certain rules it particularly important if the audience can't ask questions during the talk. There are two self-help ways to communicate your points effectively. The first is OCR, I. S. 2. (Keep it simple short). Focus on presenting one to three key points sense. Think about how much you remember from a talk last week. A presentation can easily be ruined if the content is too difficult for the audience to follow, or if the structure is too complicated. Try to with presentations simple and logical structure. Include an introduction in which you outline the points you intend to cover and a conclusion in which you go over the main points of your talk. Second, repeat key insights tell them what you're going to tell them, tell them, and tell them what you told them. (summary).

Resources for Reference

http://www.ege.ee/wiki-edu/...matduki_konferents-reife.html

Unit 05

Section 1
Intensive Reading

A City Is Not a Tree
Christopher Alexander

Part V

[*Alexander discusses Columbia, Maryland (Figure 5 – 1), and seven other city plans by famous architects and planners: Greenbelt, Maryland (Clarence Stein), Greater London (Abercrombie and Forshaw), Tokyo (Kenzo Tange), Mesa City (Paolo Soleri), Chandigarh (Le Corbusier), Brasilia (Lucio Costa), and Communitas (Paul and Percival Goodman).*]

 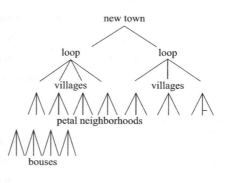

Figure 5–1 Columbia, Maryland. Community Research and Development Inc.: Neighborhoods, in cluster of live, form "village." Tranportaton joins the villages into new town. The organization is a tree.

The most beautiful example of all I have kept until last, because it symbolizes the problem perfectly. It appears in Hilberseimer's book called *The Nature of Cities*. He describes the fact that certain Roman towns had their origin as military camps, and then shows a picture of a modern military encampment as a kind of archetypal form for the city. It is not possible to have a structure which is a clearer tree.

The symbol is apt for, of course, the organization of the army was created precisely in order to create discipline and rigidity. When a city is endowed with a tree structure, this is what happens to the city and its people. Hilberseimer's own scheme for the commercial area of a city is based on the army camp archetype.

Each of these structures, then, is a tree. Each unit in each tree that I have described, moreover, is the fixed, unchanging residue of some system in the living city (just as a house is the residue of the interactions between the members of a family, their emotions, and their belongings; and a freeway is the residue of movement and commercial exchange).

However, in every city there are thousands, even millions, of times as many more systems at work whose physical residue does not appear as a unit in these tree structures. In the worst cases, the units which do appear fail to correspond to any living reality; and the real systems, whose existence actually makes the city live, have been provided with no physical receptacle.

In a traditional society, if we ask a man to name his best friends and then ask each of these in turn to name their best friends, they will all name each other so that they form a closed group. A village is made of a number of separate closed groups of this kind.

But today's social structure is utterly different. If we ask a man to name his friends and then ask them in turn to name their friends, they will all name different people, very likely unknown to the first person; these people would again name others, and so on outwards. There are virtually no closed groups of people in modern society. The reality of today's social structure is thick with overlap-the systems of friends and acquaintances form a semi-lattice, not a tree (Figure 5-2).

In the natural city, even the house on a long street (not in some little cluster) is a more accurate acknowledgement of the fact that your friends live not next door, but far away, and can only be reached by bus or automobile. In this respect Manhattan has more overlap in it than Greenbelt. And though one can argue that in Greenbelt too, friends are only minutes away by car, one must then ask: since certain groups have been emphasized by the physical units of the physical structure, why are just these the most irrelevant ones?

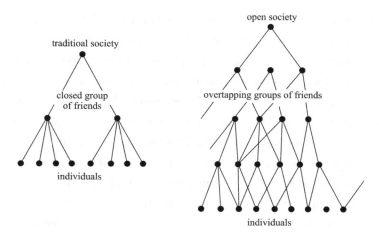

Figure 5-2

Words and Phrases

1. encampment   *n*.   营地
2. archetypal   *adj*.   原型的
3. apt   *adj*.   合适的、适当的
4. discipline   *n*.   纪律
5. rigidity   *n*.   严格
6. residue   *n*.   残留物
7. interaction   *n*.   相互作用
8. correspond to   相应、符合
9. virtually   *adv*.   事实上、实质上
10. be thick with   充满
11. acquaintance   *n*.   熟人
12. cluster   *n*.   簇、（住宅）组团
13. acknowledgement   *n*.   确认、承认
14. irrelevant   *adj*.   不相关的

Section 2
Extensive Reading

Passage 1
Greenbelt, Maryland: A Living Legacy of the New Deal
Cathy D. Knepper. (Baltimore: Johns Hopkins University Press, 2001.)

Greenbelt, Maryland, one of three communities built under the New Deal's Resettlement Administration, is a commonly celebrated case of planning ideals translated into a concrete outcome. Texts in planning and urban history cite the building of Greenbelt as a watershed in modern American urban development, focusing largely on its unique physical attributes, which combined elements of the architect Clarence Stein's aborted scheme for Radburn, New Jersey, with Clarence Perry's neighborhood unit plan, both products of 1920s reformers. At the same time, Greenbelt, like Rexford Tugwell's planned community experiments (Greenhills, Ohio and Greendale, Wisconsin), is frequently dismissed as a nice ideal that failed in implementation. As David R. Goldfield and Blaine A. Brownell note in *Urban America: A History* (1990), Tugwell's towns ended up as "little more than planned middle-class bedroom suburbs". Cathy D. Knepper's Greenbelt, Maryland, directly challenges that interpretation, noting the important set of relationships between the community's form and its future functioning. In Knepper's well-argued opinion, Greenbelt was an experiment in cooperative community building that strengthened and broadened its cooperative fabric over the next sixty years.

Passage 2
Abercrombie, Sir Patrick

1879 ~ 1957, British architect and town planner, Professor of civil design at the Univ. of Liverpool from 1915 to 1935 and of town planning at the Univ. of London after 1935, he acted as consultant in the rebuilding and planning of London, Edinburgh, Bath, and other British cities. He was knighted in 1945. His voluminous writing has been of considerable influence in the field of city and regional planning. His books include *The Preservation of Rural England* (1926) and *Town and Country Planning* (1933).

Passage 3
Mesa City

Mesa City first appears in Soleri's drawings in 1955 as "Project Mesa: quest for an environment in harmony with man". Over the next five years, Soleri would draw over a thousand feet of scrolls detailing the structures and landscape of this hypothetical city.

The Mesa City plans are generally regarded as precursors to the idea of arcology, and contain the seeds of many of Soleri's later ideas. On their own, the plans stand as a visually stunning consideration of landscape, architectural design, and large-scale urban planning.

Resources for Reference

http://www.princegeorges.org/miniwebs/Greenbelt/erin-2005/greenbelt-intro.html
http://www.arcosanti.org

Section 3
Tips for Translation

Turning words and phrases into sentences 把单词和短语翻译成句子

英语的句子结构往往比较复杂，里面包含多个短语、状语、同位语等。如果硬要按照原文对译，会使句子臃肿、晦涩。在这种情况下，把某些句子成分转换成相对独立的短句翻译，反而能更加准确、完整地表达原意。

1. The entrance space is a Beaux-Arts hall *with a richly decorated staircase*, *divided by groups of sofas into several seating areas.*
入口空间是一个学院派风格的大厅，里面有一部装饰丰富的楼梯，并被成组的沙发分隔成几个休息区。
如果硬把上句中的介词短语和过去分词短语译成"hall"的定语，反倒会使译文头重脚轻。

2. *Surprisingly enough*, the students managed to submit theirs works before the deadline in the end.
令人惊讶的是，同学们居然在截止日期前交出了作业。
状语（Surprisingly enough）译成句子

3. The Peace Hotel, *one of the historical monuments of Shanghai*, saw the rise of the city at the turn of the 19th and 20th century.
和平饭店是上海历史保护建筑之一，它见证了19、20世纪之交上海的崛起。
同位语（one of the historical monuments of Shanghai）译成句子。

4. Alexander is a self-proclaimed iconoclast, deliberately distancing himself from virtually all the major mainstream currents of twentieth-century architectural and planning thought.
亚历山大自称是一个破坏偶像的人，他有意识地同20世纪建筑学和城市规划的主流思想保持距离。
现在分词短语（deliberately distancing himself…thought.）译成句子。

Section 3
Tips for Translation

Turning words and phrases into sentences by adding words or ideas

1. The entrance space is a 26-meter-wide hall with a richly decorated staircase, divided in 1 concave rows into seated theater area.

2. Surprisingly enough, the student managed to submit three papers before the due date in time at the end.

3. The Peace Hotel, one of the cherished monuments of Shanghai, marks the rise of the city at the turn of the 19th and 20th century.

4. Alex adopted a self-proclaimed avant-last, deliberately detaching himself from virtually of the major mainstream currents of twentieth-century architectural and planning thought.

Unit 06

Section 1
Intensive Reading

A City Is Not a Tree
Christopher Alexander

Part VI

In the second part of this paper, I shall further demonstrate why the living city cannot be properly contained in a receptacle which is a tree-that indeed, its very life stems from the fact that it is not a tree.

Finally, I shall try to show that it is the process of thought itself which works in a treelike way, so that whenever a city is "thought out" instead of "grown", it is bound to get a treelike structure.

In the first part of this article, we saw that the units of which an artificial city is made up are organized to form a tree. So that we get a really clear understanding of what this means, and shall better see its implications, let us define a tree once again.

Whenever we have a tree structure, it means that within this structure no piece of any unit is ever connected to other units, except through the medium of that unit as a whole.

The enormity of this restriction is difficult to grasp. It is a little as though the members of a family were not free to make friends outside the family, except when the family as a whole made a friendship.

In simplicity of structure the tree is comparable to the compulsive desire for neatness and order that insists the candlesticks on a mantelpiece be perfectly straight and perfectly symmetrical about the center. The semi-lattice, by comparison, is the structure of a complex fabric; it is the structure of living things; of great paintings and symphonies.

It must be emphasized, lest the orderly mind shrink in horror from anything that is not clearly articulated and categorized in tree form, that the idea of overlap, ambiguity, multiplicity of aspect, and the semi-lattice, are not less orderly than the rigid tree, but more so. They represent a thicker, tougher, more subtle and more complex view of structure.

Let us now look at the ways in which the natural, when unconstrained by artificial conceptions, shows itself to be a semi-lattice.

A major aspect of the city's social structure which a tree can never mirror properly is illustrated by Ruth Glass's redevelopment plan for Middlesborough, a city of 200,000 which she recommends be broken down into 29 separate neighborhoods. After picking her 29 neighborhoods by determining where the sharpest discontinuities of building type, income, and job type occur, she asks herself the question: "If we examine some of the social systems which actually exist for the people in such a neighborhood, do the physical units defined by these various social systems all define the same spatial neighborhood?" Her own answer to this question is no.

Each of the social systems she examines is a nodal system. It is made of some sort of central node, plus the people who use this center. Specifically she takes elementary schools, secondary schools, youth clubs, adult clubs, post offices, greengrocers, and grocers selling sugar. Each of these centers draws its users from a certain spatial area or spatial unit. This spatial unit is the physical residue of the social system as a whole, and is therefore a unit in the terms of this paper. The units corresponding to different kinds of centers for the single neighborhood of Waterloo Road are shown in Figure 6-1. The hard outline is the boundary of the so-called neighborhood itself. The white circle stands for the youth club, and the small solid rings stand for areas where its members live. The ringed spot is the adult club, and the homes of its members form the unit marked by dashed boundaries. The white square is the post office and the dotted line marks the unit which contains its users. The secondary school is marked by the spot with a black triangle in it. Together with its pupils, it forms the system marked by the dot-dashed line.

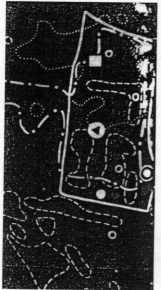

Figure 6-1

As you can see at once, the different units do not coincide. Yet neither are they disjoint. They overlap.

Words and Phrases

1. stem   *v.*   生长
2. be bound to   一定要
3. implication   *n.*   含义、暗示
4. enormity   *n.*   巨大
5. candlestick   *n.*   烛台
6. mantelpiece   *n.*   壁炉架
7. symmetrical   *adj.*   对称的
8. symphony   *n*   交响乐
9. lest   *conj.*   以免
10. articulate   *v.*   说清楚、接合
11. categorize   *v.*   分类
12. ambiguity   *n.*   含糊
13. multiplicity   *n.*   多样性
14. nodal   *adj.*   节（点）的
15. coincide   *v.*   一致

Section 2
Extensive Reading

Sir Raymond Unwin

Sir Raymond (1863 ~ 1940) was an influential town planner and an authority on the design and management of garden cities. Together with his partner Barry Parker (1867 ~ 1941), he designed Letchworth Garden City in 1903 and Hampstead Garden Suburb in 1907.

He is one of the most important figures in town planning and housing standards, who could combined his technical skills and grasp of sociological issues with his ability to communicate his ideas. He loved pure clean design, and would go out of his way not to destroy natural beauty, but remained equally aware of the needs of economics, or social psychology.

1909-In response to the pressure by Unwin and others to bring in legislation to bring the new residential standards into more general use the First Housing and Town Planning Act was passed. This enabled local authorities to regulate suburban development's layout and density.

One of the first schemes prepared was for Ruislip Northwood. Thomas Adams was Principal Planning Inspector at the Local Government Board and Unwin was a consultant to the principal owner of the land.

Unwin publishes "Town Planning in Practice" as a text book reviewing historic towns and residential development and putting forward his arguments on housing density and layout.

Resources for Reference

http://www.letchworthgardencity.net

Section 3
Tips for Translation

Translating personal and possessive pronouns 人称代词和物主代词的翻译

在英语中大量采用人称代词或物主代词取代前面出现过的名词或名词短语，以避免重复。但在汉译中，往往需要还原成人称或物主代词所取代的名词或短语，使翻译明确、妥帖。

1. The most beautiful example of all I have kept until last, because it symbolizes the problem perfectly. *It* appears in Hilberseimer's book called The Nature of Cities.

我把最好的例子保留到最后，因为它最完美地说明了问题。这一例子出现在希伯塞莫《城市的本质》一书中。

后面一句句子中的"it"如果翻译成"它"，就会显得不那么自然，因此在翻译的时候不如还原成它指代的对象"example"。

2. Sir Raymond was an influential town planner and an authority on the design and management of garden cities. Together with his partner Barry Parker, *he* designed Letchworth Garden City in 1903 and Hampstead Garden Suburb in 1907.

雷蒙德爵士是有影响力的田园城市设计和管理方面的规划师和权威。雷蒙德爵士同他的合作者巴里·帕克一起，1903 设计了莱彻沃斯田园城，1907 年设计了罕普斯泰德田园郊区。

后面一句中的"he"可以还原成"雷蒙德爵士"，避免同紧接着的"他的"重复。

3. Alexander's approach to describing how cities should be designed in this selection may trouble readers who seek clear, rational guidelines. *He* takes the position that not enough is yet known about how to design non-treelike cities to provide definite answers.

亚历山大在文章节选中提出的城市设计策略可能会困扰那些寻求清晰、理性设计准则的读者。亚历山大的观点是：关与如何设计非树状结构的城市还所知不够，从而无法提出明确的解决方案。

后面一句中的"he"可以还原成"亚历山大"，使翻译明确、清晰。

Unit 07

Section 1
Intensive Reading

A City Is Not a Tree
Christopher Alexander

Part VII

We cannot get an adequate picture of what Middlesborough is, or of what it ought to be, in terms of 29 large and conveniently integral chunks called neighborhoods. When we describe the city in terms of neighborhoods, we implicitly assume that the smaller elements within any one of these neighborhoods belong together so tightly that they interact with elements in other neighborhoods only through the medium of the neighborhood to which they themselves belong. Ruth Glass herself shows clearly that this is not the case.

Below (Figure 7 – 1) are two pictures of the Waterloo neighborhood. For the sake of argument, I have broken it into a number of small areas. The left-hand diagram shows how these pieces stick together in fact, and the right-hand diagram shows how the redevelopment plan pretends they stick together.

There is nothing in the nature of the various centers which says that their catchment areas should be the same. Their natures are different. Therefore the units they define are different. The natural city of Middlesborough was faithful to the semi-lattice structure they have. Only in the artificial tree conception of the city are their natural, proper, and necessary overlaps destroyed.

Take the separation of pedestrians from moving vehicles, a tree concept proposed by Le Corbusier,

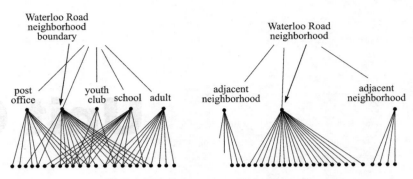

Figure 7-1　Waterloo neighborhood.

Louis Kahn, and many others. At a very crude level of thought this is obviously a good idea. It is dangerous to have 60-mile-an-hour cars in contact with little children toddling. But it is not always a good idea. There are times when the ecology of a situation actually demands the opposite. Imagine yourself coming out of a Fifth Avenue store; you have been shopping all afternoon; your arms are full of parcels; you need a drink; your wife is limping. Thank God for taxis.

Yet the urban taxi can function only because pedestrians and vehicles are not strictly separated. The prowling taxi needs a fast stream of traffic so that it can cover a large area to be sure of finding a passenger. The pedestrian needs to be able to hail the taxi from any point in the pedestrian world, and to be able to get out to any part of the pedestrian world to which he wants to go. The system which contains the taxicabs needs to overlap both the fast vehicular traffic system and the system of pedestrian circulation. In Manhattan pedestrians and vehicles do share certain parts of the city, and the necessary overlap is guaranteed.

Another favorite concept of the CIAM theorists and others is the separation of recreation from everything else. This has crystallized our real cities in the form of playgrounds. The playground, asphalted and fenced in, is nothing but a pictorial acknowledgment of the fact that "play" exists as an isolated concept in our minds. It has nothing to do with the life of play itself. Few self-respecting children will even play in a playground.

Play itself, the play that children practice, goes on somewhere different every day. One day it may be indoors, another day in a friendly gas station, another day down by the river, another day in 2 derelict building, another day on a construction site which has been abandoned for the weekend. Each of these play activities, and the objects it requires, forms a system. It is not true that these systems exist in isolation, cut off from the other systems in the city. The different systems overlap one another, and they overlap many other systems besides. The units, the physical places recognized as play places, must do the same.

In a natural city this is what happens. Play takes place in a thousand places-it fills the interstices of adult life. As they play, children become full of their surroundings. How can a child become filled with his surroundings in a fenced enclosure? It cannot.

A similar kind of mistake occurs in trees like that of Goodman's Communitas, or Soleri's Mesa City, which separate the university from the rest of the city. Again, this has actually been realized in common American form of the isolated campus.

What is the reason for drawing a line in the city so that everything within the boundary is university, and everything outside is non-university? It is conceptually clear. But does it correspond to the realities of university life? Certainly it is not the structure which occurs in non-artificial university cities.

Take Cambridge University for instance. At certain poises Trinity Street is physically almost indistinguishable from Trinity College. One pedestrian crossover in the street is literally part of the college. The buildings on the street, though they contain stores and coffee shops and banks at ground level, contain undergraduates' rooms in their upper stories. In many cases the actual fabric of the street buildings melts into the fabric of the old college buildings.

There will always be many systems of activity where university life and city life overlap: pub-crawling, coffee-drinking, the movies, waving from place to place. In some cases whole departments may be actively involved in the life of the city's inhabitants (the hospital-cum-medical school is an example). In Cambridge, a natural city where university and city have grown together gradually, the physical units overlap because they are the physical residues of city systems and university systems which overlap (Figure 7-2).

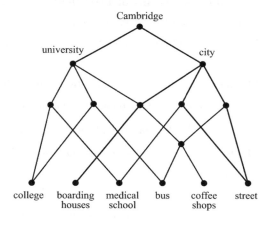

Figure 7-2

Words and Phrases

1. implicitly *adv.*   含蓄地、暗中地
2. redevelopment *n.*   再开发
3. catchment *n.*   集水处、流域
4. toddle *v.*   东倒西歪地走路、蹒跚学步
5. limp *v.*   跛行
6. prowl *v.*   巡游、徘徊
7. hail *v.*   打招呼

8. pedestrian circulation 步行流线
9. recreation *n.* 娱乐
10. crystallize *v.* 明确
11. asphalt *n.* 沥青、柏油
12. interstice *n.* 裂缝
13. crossover *n.* 天桥
14. literally *adv.* 差不多
15. melt into 融入
16. pub-crawl *v.* 逛酒店喝酒
17. inhabitant *n.* 居民
18. hospital-cum-medical school *n.* 医院附属医学院

Section 2
Extensive Reading

Passage 1
Communitas, Means of Livelihood and Ways of Life
Percival Goodman and Paul Goodman
Preface by Paul Goldberger

"Communitas stands in a class by itself: a fresh and original theoretic contribution to the art of building cities. Such a book does not appear often…a witty, penetrating, provocative and, above all…a wise book; for it deals with the underlying values and purposes, political and moral, on which planning of any sort must be based."
—Lewis Mumford

"Rich in splendid observations, many of which foreshadow issues which have become all the more urgent today."
—Paul Goldberger

Passage 2
A Brief Biography of Lewis Mumford (1895~1990)
by Eugene Halton

Internationally renowned for his writings on cities, architecture, technology, literature, and modern life, Lewis Mumford was called by Malcolm Cowley, "the last of the great humanists". His contributions to literary criticism, architectural criticism, American studies, the history of cities, civilization, and technology, as well as to regional planning, environmentalism, and public life in America, mark him as one of the most original voices of the twentieth-century.

Born in Flushing on October 19, 1895, Mumford lived much of his life in New York, settling in Dutchess County in 1936 with his wife Sophia, in Amenia, where he died over a half-century later, on January 26, 1990. His first book, *The Story of Utopias*, was published in 1922, and his last book, his autobiography *Sketches from Life*, was published sixty years later in 1982.

Mumford preferred to call himself a writer, not a scholar, architectural critic, historian or philosopher. His writing ranged freely and brought him into contact with a wide variety of people, including writers, artists, city planners, architects, philosophers, historians, and archaeologists. Throughout his life, Mumford sketched and painted his surroundings, visualizing his impressions of people and places in image, as his ever-present notepad visualized them in words.

Given the range of Mumford's scholarly work, it is all the more interesting that he did not have a college degree, having had to leave City College of New York after a diagnosis of tuberculosis. But if whaling was Herman Melville's "Harvard and Yale", "Mannahatta", as Mumford put it, "was my university, my true alma mater". From childhood on, Mumford walked, sketched, and observed New York City, and its effects can be felt throughout his writings.

He was architectural critic for the New Yorker magazine for over thirty years, and his 1961 book, *The City in History*, received the National Book Award. In 1923 Mumford was a cofounder with Clarence Stein, Benton MacKaye, Henry Wright and others, of the Regional Planning Association of America, which advocated limited-scale development and the region as significant for city planning.

By 1938 he was an ardent advocate for early American entry into what was emerging as World War Two, a war which claimed the life of his son Geddes in 1944, and was an early critic of nuclear weapons in 1946 and of U.S. involvement in Vietnam in 1965. In 1964 he was awarded the Presidential Medal of Freedom.

Lewis Mumford's work underwent a continuous series of transformations as he broadened and deepened his scope. From his American studies books in the 1920s-such as *The Golden Day* (1926) and *Herman Melville* (1929), which contributed to the rediscovery of the literary transcendentalists of the 1850s, and *The Brown Decades* (1931) which placed the architectural achievements of Henry Hobson Richardson, Louis Sullivan and Frank Lloyd Wright before the public, through the four-volume "Renewal of Life" series published between 1934 and 1951, which outlined the place of technics, cities, and world-views in the development of Western Civilization, to his late studies of the emergence of civilizations and the place of communication practices in human development, he boldly denied the utilitarian view while evolving his own vision of organic humanism.

Mumford's works share a common concern with the ways that modern life as a whole, although providing possibilities for broader expression and development, simultaneously subverts those possibilities and actually ends up tending toward a diminution of purpose. He shows in lucid detail how the modern ethos released a Pandora's box of mechanical marvels which eventually threatened to absorb all human purposes into *The Myth of the Machine*, the title he used for his two-volume late work.

See, for example, Lewis Mumford's critique of the World Trade Center from 1970, when it was just being built.

Despite what he saw as a likelihood of catastrophic dehumanization, he argued for the hope that the organic depths of human nature, of the "fibrous structure of history", might provide the basis for a transformation of megatechnic civilization.

Mumford argued passionately for a restoration of organic human purpose in the larger scheme of things, a task requiring a human personality capable of "primacy over its biological needs and technological pressures", and able to "draw freely on the compost from many previous cultures".

As he wrote in his 1946 book, Values for Survival:
"If we are to create balanced human beings, capable of entering into world-wide co-operation with all other men of good will——and that is the supreme task of our generation, and the foundation of all its other potential achievements——we must give as much weight to the arousal of the emotions and to the expression of moral and esthetic values as we now give to science, to invention, to practical organization. One without the other is impotent. And values do not come ready-made: they are achieved by a resolute attempt to square the facts of one's own experience with the historic patterns formed in the past by those who devoted their whole lives to achieving and expressing values. If we are to express the love in our own hearts, we must also understand what love meant to Socrates and Saint Francis, to Dante and Shakespeare, to Emily Dickinson and Christina Rossetti, to the explorer Shackleton and to the intrepid physicians who deliberately exposed themselves to yellow fever. These historic manifestations of love are not recorded in the day's newspaper or the current radio program: they are hidden to people who possess only fashionable minds. Virtue is not a chemical product, as Taine once described it: it is a historic product, like language and literature; and this means that if we cease to care about it, cease to cultivate it, cease to transmit its funded values, a large part of it will become meaningless, like a dead language to which we have

lost the key. That, I submit, is what has happened in our own lifetime."

Mumford's 1982 autobiography was followed by a biography by Donald Miller in 1989, *Lewis Mumford: A Life*. Mumford's papers are stored in Philadelphia at the Van Pelt Library of The University of Pennsylvania, and his library and watercolors and drawings are stored at the library of Monmouth University, West Long Branch, New Jersey.

Resources for Reference

http://en.wikipedia.org/wiki/Communitas
http://www.nd.edu/~ehalton/mumfordbio.html

Section 3
Tips for Translation

Amplification 加词

在英译汉中，根据上下文以及汉语的语言习惯，把句子中省略的成分或者隐含的意思补充完整，使译文准确、完整。还有一种比较常见的情况是在翻译英语中的抽象名词（abstract noun）的时候，通过适当增补搭配词，是翻译更符合汉语的习惯。

1. In Vienna, London, Paris, each of the performing arts has found its own place, because *all* are not mixed randomly. *Each* has created its own familiar section of the city.
在维也纳、伦敦和巴黎，每种表演艺术都有其固有的场所，因为不是所有的表演艺术都随随便便地被混杂在一起的。每种表演艺术在城市之中都有其为人所熟知的所在。
上文中"all"和"each"都指代的是"performing art"，在翻译中宜把"表演艺术"补充进去，使翻译清楚、顺畅。

2. The total separation of work from housing, started by Tony Gamier in his industrial city, then incorporated in the 1929 Athens Charter, is now found in every artificial city and accepted everywhere where zoning is enforced. ⋯ But the *separation* misses a variety of systems which require, for their sustenance, little parts of *both*.
最早由托尼·伽涅在其工业城市概念中提出的工作和居住功能的绝对分离后来成为1929年《雅典宪章》的组成部分，如今每个人工城市都采取了这一做法，并由于控制性规划的强制性要求而得到了广泛地接受。但工作和居住的分离忽略了这样一种情况：多样化的城市系统必须往往必须同时具有工作和居住的一小部分功能才能存在。
上文的后一句中的"separation"指的是前一句中的"the total separation of work from housing"，因此在翻译中应把这一意思补充进去，译为"工作和居住功能的分离"。"both"指的是"工作和居住"两种功能，在翻译中也应当加上。

3. The community where the worker lives, if it is mainly residential, collects only little in the way of taxes, and yet has great additional burdens on its purse in the shape of *schools*, *hospitals*, etc.
工人社区如果主要以居住功能为主的话,税收就会非常少,在学校和医院建设方面就会承担很大的资金压力。

试比较"在学校和医院方面就会承担很大的资金压力"和"在学校和医院建设方面就会承担很大的资金压力"两句译文,显然后一句翻译更为明确。

4. The students were anxious to relax with the weeks of *strains* before their design were submitted relieved.
学生们渴望从交图前几个星期的紧张状态中解脱出来。"strain"在此翻译成"紧张状态"较为顺畅。

5. *Preparation* for the exhibition of the international students' works of Shanghai Biennale continued.
上海双年展国际学生展览的准备工作继续进行。
"preparation"是由动词派生出来的抽象名词,这一类词往往在翻译时需要加词。

Unit 08

Section 1
Intensive Reading

A City Is Not a Tree
Christopher Alexander

Part VIII

Let us look next at the hierarchy of urban cores, realized in Brasilia, Chandigarh, the MARS plan for London, and, most recently, in the Manhattan Lincoln Center, where various performing arts serving the population of greater New York have been gathered together to form just one core.

Does a concert hall ask to be next to an Opera House? Can the two feed on one another? Will anybody ever visit them both, gluttonously, in a single evening, or even buy tickets from one after going to a concert in the other? In Vienna, London, Paris, each of the performing arts has found its own place, because all are not mixed randomly. Each has created its own familiar section of the city. In Manhattan itself, Carnegie Hall and the Metropolitan Opera House were not built side by side. Each found its own place, and now creates its own atmosphere. The influence of each overlaps the parts of the city which have been made unique to it.

The only reason that these functions have all been brought together in the Lincoln Center is that the concept of performing art links them to one another.

But this tree, and the idea of a single hierarchy of urban cores which is its parent, do not illuminate the relations between art and city life. They are merely born of the mania every simpleminded person has for putting things with the same name into the same basket.

The total separation of work from housing, started by Tony Garnier in his industrial city, then incorporated in the 1929 Athens Charter, is now found in every artificial city and accepted everywhere where zoning is enforced. Is this a sound principle? It is easy to see how bad conditions at the beginning of the century prompted planners to try to get the dirty factories out of residential areas. But the separation misses a variety of systems which require, for their sustenance, little parts of both.

Jane Jacobs describes the growth of backyard industries in Brooklyn. A man who wants to start a small business needs space, which he is very likely to have in his own backyard. He also needs to establish connections with larger going enterprises and with their customers. This means that the system of backyard industry needs to belong both to the residential zone, and to the industrial zone——these zones need to overlap. In Brooklyn they do. In a city which is a tree, they can't.

Finally, let us examine the subdivision of the city into isolated communities. As we have seen in the Abercrombie plan for London, this is itself a tree structure. The individual community in a greater city has no reality as a functioning unit. In London, as in any great city, almost no one manages to find work which suits him near his home. People in one community work in a factory which is very likely to be in another community.

There are, therefore, many hundreds of thousands of worker-workplace systems, each consisting of a man plus the factory he works in, which cut across the boundaries defined by Abercrombie's tree. The existence of these units, and their overlapping nature, indicates that the living systems of London form a semi-lattice. Only in the planner's mind has it become a tree.

The fact that we have so far failed to give this any physical expression has a vital consequence. As things are, whenever the worker and his workplace belong to separately administered municipalities, the community which contains the workplace collects huge taxes and has relatively little on which to spend the tax revenue. The community where the worker lives, if it is mainly residential, collects only little in the way of taxes, and yet has great additional burdens on its purse in the shape of schools, hospitals, etc. Clearly, to resolve this inequity, the worker-workplace systems must be anchored in physically recognizable units of the city which then can be taxed.

Now, why is it that so many designers have conceived cities as trees when the natural structure is in every case a semi-lattice? Have they done so deliberately, in the belief that a tree structure will serve the people of the city better? Or have they done it because they cannot help it, because they are trapped by a mental habit, perhaps even trapped by the way the mind works; because they cannot encompass the complexity of semi-lattice in any convenient mental form; because the mind has an overwhelming predisposition to see trees wherever it looks and cannot escape the tree conception?

I shall try to convince you that it is for this second reason that trees are being proposed and built as cities——that it is because designers, limited as they must be by the capacity of the mind in form intuitively accessible structures, cannot achieve the complexity of the semi-lattice in a single mental act.

Words and Phrases

1. gluttonously   *adv.*   贪婪地、贪吃地
2. Carnegie Hal   （美）卡耐基音乐厅
3. Metropolitan Opera House   （美）大都会歌剧院
4. Unique   *adj.*   独一无二的
5. performing art   表演艺术
6. hierarchy   *n.*   层级、层次
7. illuminate   *v.*   阐明、照亮
8. mania   *n.*   狂热
9. simpleminded   *adj.*   头脑简单的
10. Athens Charter   雅典宪章
11. zonin   *n.*   分区
12. prompt   *v.*   促使
13. sustenance   *n.*   生存、生计、维持
14. enterprise   *n.*   企业
15. cut across   抄近路
16. vital   *adj.*   至关重要的
17. consequence   *n.*   结果
18. municipality   *n.*   市政当局
19. revenue   *n.*   税收
20. in the shape of   以……的形式
21. inequity   *n.*   不公平、不公正
22. anchor   *v.*   锚固
23. deliberately   *adv.*   故意地
24. encompass   *v.*   包含、环绕
25. overwhelming   *adj.*   压倒性的
26. predisposition   *n.*   素质、倾向
27. intuitively   *adv.*   直觉地
28. accessible   *adj.*   可达的

Section 2
Extensive Reading

Passage 1
Jane Jacobs

Jane Jacobs (Figure 8-1) has no professional training in the field of city planning, nor does she

hold the title of urban planner anywhere. However, she has used her own observations about cities to formulate her philosophy about them. Though some of her views go against the traditional views on planning, her work is well respected by practicing planners and planning students alike.

It is obvious that Jane Jacobs is a woman with strong ideas who is not afraid to share them, even if they may not necessarily agree with the ideas of others in her field of interest. Her original ideas give us the opportunity to look at cities in a different light. It is our hope that through the projects presented on this page, everyone can get a sense of what these ideas are and use them in forming their own way of looking at cities.

Books by Jane Jacobs:

Figure 8-1   Jane Jacobs

The Death and Life of Great American Cities, 1961.
The Economy of Cities, 1969.
A Question of Separatism: Quebec and the Struggle over Sovereignty, 1980.
Cities and the Wealth of Nations, 1984.
Systems of Survival: A Dialogue on the Moral Foundations of Commerce and Politics, 1992.

Passage 2
Philip Johnson

Philip Johnson (Figure 8-2), whose austere "glass box" buildings and latter-day penchant for incorporating whimsical touches in his designs made him one of the most influential architects of the 20th century, has died at 98.

Johnson died Tuesday at his home in New Canaan, Conn.——itself one of his most important creations.

Johnson's work, which spanned more than half a century starting in the 1940s, ranged from the modernism of his home, a glass cube in the woods, to the more fanciful work of his later years, including the AT&T Building in New York, with its curved pediment that made it look like a giant Chippendale chest of drawers.

Figure 8-2   Philip Johnson

Johnson once said his great ambition was "to build the greatest room in the world——a great theater or cathedral or monument. Nobody's given me the job".

In 1980, however, he completed his great room, the Crystal Cathedral in Garden Grove, Calif., a soaring glass structure wider and higher than Notre Dame in Paris. If architects are remembered for their one-room buildings, Johnson said, "This may be it for me."

With his partner, John Burgee, Johnson also designed the Bank of America building in Houston, a 56-story tower of pink granite stepped back in a series of Dutch gable roofs; and the Cleveland Playhouse, a complex with the feel of an 11th-century town.

"The world has lost a towering force who defined the art and practice of architecture in the 20th century," said architect Daniel Libeskind, the master planner for the new towers rising on the site of the World Trade Center.

Johnson was one of architecture's most recognizable figures, with his trademark black round-rimmed glasses that gave him an owlish look.

Pulitzer Prize-winning architecture critic Paul Goldberger of The New Yorker pronounced him "the greatest architectural presence of our time——which is not the same thing as the greatest architect".

"He was probably our first and most significant architect as celebrity," Goldberger said. "There's no question that he used his fame for the betterment of architecture. His greatest passion was in seeing architecture, talking about it, making a stimulating dialogue about it."

Johnson also invented the role of museum architecture curator, at New York's Museum of Modern Art in 1932. And he coined the term International Style for the work of Europeans Ludwig Mies van der Rohe and Le Corbusier.

His efforts to bring their style to the United States and incorporate some of its elements in his own work "literally changed the landscape of American architecture", said Terrence Riley, MoMA's current Philip Johnson chief curator for architecture and design.

Johnson got a chance to work with van der Rohe by designing the interiors for the German architect's Seagram Building in New York.

He died this week with his latest project in the works: an "urban glass house" in New York's Soho neighborhood, inspired by his home in New Canaan. Until last year, the architect came to his office three days a week, said Alan Ritchie, Johnson's partner at Philip Johnson-Alan Ritchie Architects in New York for three decades.

Johnson's AT&T Building, a granite-walled tower with an enormous 90-foot arched entryway and a fanciful top, broke decisively with the glass towers that crowded Manhattan. The building, completed in 1983 and now owned by Sony, marked a sharp turn in architectural taste away from the clean lines of modernism. Other architects felt emboldened to experiment with styles, and commissions poured into the offices of Johnson and Burgee.

Most of the firm's projects were corporate palaces: the Transco II and Bank of America towers in Houston, in 1983 and 1984; a 23-story neo-Victorian office building in San Francisco; and a mock-gothic glass tower for PPG Industries in Pittsburgh, built in 1983.

"The people with money to build today are corporations——they are our popes and Medicis," Johnson said. "The sense of pride is why they build."

Toward the end of his life, Johnson went public with some private matters——his homosexuality and his past as a disciple of Hitler-style fascism. On the latter, he said he spent much time in Berlin in the 1930s and became "fascinated with power", but added he did not consider that an excuse.

"I have no excuse (for) such utter, unbelievable stupidity…. I don't know how you expiate guilt". he said.

He said that it was his homosexuality that caused him to suffer a nervous breakdown while he was a student at Harvard. He said that in 1977 he asked The New Yorker magazine to omit references to it, fearing he might lose the AT&T commission, which he called "the job of my life".

Philip Cortelyou Johnson was born July 8, 1906, in Cleveland, the only son of lawyer Homer H. Johnson and his wife, Louise. He graduated from Harvard in 1927 with a degree in philosophy, then toured Europe and became interested in new styles of architecture.

In 1940, Johnson returned to Harvard for graduate school, studying under Marcel Breuer. He returned to MoMA, then left in 1955 to open his own design office.

His projects at times ran into criticism from preservationists and even fellow architects. In 1987, he was replaced as designer of the second phase of the New England Life Insurance Co. headquarters in Boston after residents complained about the project's size and style.

Critics unearthed a quotation he had made at a conference a couple of years earlier: "I am a whore and I am paid very well for high-rise buildings." Johnson said later that his choice of words was unfortunate and he meant only that architects need to be able to compromise with developers.

In 1979, Johnson became the first recipient of the prestigious Pritzker Architecture Prize.

Early in his career, he reflected on what he had hoped to achieve.

"I like the thought that what we are to do on this earth is embellish it for its greater beauty," he said, "so that oncoming generations can look back to the shapes we leave here and get the same thrill that I get in looking back at theirs——at the Parthenon, at Chartres Cathedral."

ON THE NET
Pritzker Prize in architecture: http://www.pritzkerprize.com
Johnson's design firm: http://www.pjar.com
(FROM San Jose Mercury News)

Resources for Reference

http://bss.sfsu.edu/pamuk/urban/bio.html
http://www.encyclopedia.com/doc/1P1-104828612.html

Section 3
Tips for Oral Presentation

Speech preparation as a process

You have to gather facts and arrange your thoughts. As you collect the ideas, you have to nurture your ideas and think about a unique way to express them in an organized manner.
A speech needs time to grow. Prepare for weeks, sleep on it, dream about it and let your ideas sink into your subconscious. Ask yourself questions, write down your thoughts, and keep adding new ideas. As you prepare every speech ask yourself the following questions.

- In one concise sentence, what is the purpose of this speech?
- Who is the audience? What is their main interest in this topic?
- What do I really know and believe about this topic as it relates to this audience?
- What additional research can I do?
- What are the main points of this presentation?
- What supporting information and stories can I use to support each of my main points?
- What visual aids, if any, do I need?
- Do I have an effective opening grabber?

- In my final summary, how will I plan to tell them "What's In It For Me?"
- Have I polished and prepared the language and words I will use?
- Have I prepared a written and concise introduction for myself?
- Have I taken care of the little details that will help me speak more confidently?

Resource for Reference

(http://www.ljlseminars.com/speech.htm)

Section 4
Tips for Writing-Principles of Writing for Impact (Part I)

Write as you would talk

There's an odd but widely held perception that writing is different from speech, that a certain formality is required that differentiates writing from speech. The perception that writing is somehow different from talking is more at the root of pompous, hard-to-grasp language than nearly any other cause.

Compare the two examples from project correspondence below:

(a) Implementation of the schematic design phase will be initiated as soon as proper authorization is received.

(b) We will begin schematic design as soon as you authorize us.

No professional would talk like example (a), which you have to read twice before the sun of meaning pokes through the clouds. Example (b) is therefore suggested.

Unit 09

Section 1
Intensive Reading

A City Is Not a Tree
Christopher Alexander

Part IX

Let me begin with an example.

Suppose I ask you to remember the following four objects: an orange, a watermelon, a football, and a tennis ball. How will you keep them in your mind, in your mind's eyes? However you do it, you will do it by grouping them. Some of you will take the two fruits together, the orange and the watermelon, and the two sports balls together, the football and the tennis ball. Those of you who tend to think in terms of physical shape may group them differently, taking the two small spheres together——the orange and the tennis ball——and the two larger and more egg-shaped objects——the watermelon and the football. Some of you will be aware of both.

Let us make a diagram of these groupings (Figure 9-1).

Either grouping taken by itself is a tree structure. The two together are a semi-lattice.

Now let us try to visualize these groupings in the mind's eye. I think you will find that you cannot visualize all four sets simultaneously——because they overlap. You can visualize one pair of sets and then the other, and you can alternate between the two pairs extremely fast, so fast that you may deceive yourself into thinking you can visualize them all together. But in truth, you cannot

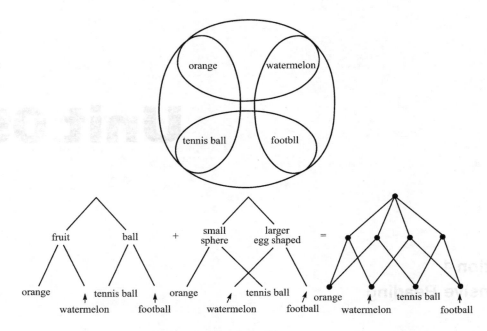

Figure 9-1

conceive all four sets at once in a single mental act. You cannot bring the semi-lattice structure into a visualizable form for a single mental act. In a single mental act you can only visualize a tree.

This is the problem we face as designers. While we are not, perhaps, necessarily occupied with the problem of total visualization as asingle mental act the principle is still the same. The tree is accessible mentally, and easy to deal with. The semi-lattice is hard to keep before the mind's eye, and therefore hard lo deal with.

It is known today that grouping and categorization are among the most primitive psychological processes. Modem psychology treats thought as a process of fitting new situations into existing slots and pigeonholes in the mind. Just as you cannot put a physical thing into more than one physical pigeonhole at once, so, by analogy, the processes of thought prevent you from putting a mental construct into more than one mental category at once. Study of the origin of these processes suggests that they stem essentially from the organism's need to read the complexity of itsenvironment by establishing barriers between the different events which it encounters.

It is for this reason that the mind's first function is to reduce the ambiguity and overlap in a confusing situation, and because to this end it is endowed with a basic for ambiguity——that structures like the city, which do require overlapping sets within them intolerance, are nevertheless persistently conceived as trees.

You are no doubt wondering, by now, what a city looks like which is a semi-lattice, but not a tree. I must confess that I cannot yet show you plans or sketches. It is not enough merely to make a demonstration of overlap——the overlap must be the right overlap. This is doubly important, because it is so tempting to make plans in which overlap occurs for its own sake. This is essentially what the high-density "life-filled" city plans of recent years do. But overlap alone does not give structure. It can also give chaos. A garbage can is full of overlap. To have structure, you must have the right overlap, and this is for us almost certainly different from the old overlap which we observe in historic cities. As the relationships between functions change, so the systems which need to overlap in order to receive these relationships muse also change. The re-creation of old kinds of overlap will be inappropriate, and chaotic instead of structured.

The work of trying to understand just what overlap the modern city requires, and trying to put this required overlap into physical and plastic terms, is still going on. Until the work is complete, there is no point in presenting facile sketches of ill thought out structure.

However, I can perhaps make the physical consequences of overlap more comprehensible by means of an image. The painting illustrated is a work by Simon Nicholson (Figure 9–2). The fascination of this painting lies in the fact that although it is constructed of rather few simple triangular elements, these elements unite in many different ways to form the larger units of the painting-in such a way, indeed, that if we make a complete inventory of the perceived units in the painting, we find that each triangle enters into

Figure 9–2

four or five completely different kinds of unit, none contained in the others, yet all overlapping in that triangle.

Thus, if we number the triangles and pick out the sets of triangles which appear as strong visual units, we get the semi-lattice shown in Figure 9–3.

[Triangles] 3 and 5 form a unit because they work together as a rectangle; 2 and 4 because they form a parallelogram; 5 and 6 because they are both dark and pointing the same way; 6 and 7 because one is the ghost of the other shifted sideways; 4 and 7 because they are symmetrical with one another; 4 and 6 because they form another rectangle; 4 and 5 because they form a sort of Z; 2 and 3 because they form a rather thinner kind of Z; 1 and 7 because they are at opposite corners; 1 and 2 because they are a rectangle; 3 and 4 because they point the same way as 5 and 6, and form a sort of off-center reflection; 3 and 6 because they enclose 4 and 5; 1 and 5 because they enclose 2, 3, and 4. I have only listed the units of two triangles. The larger units are even more complex. The white is more complex still, and is not even included in the diagram because it is harder to be sure of its elementary pieces.

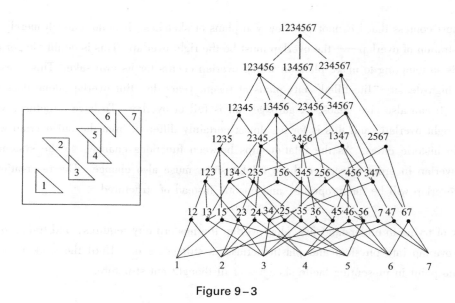

Figure 9-3

The painting is significant, not so much because it has overlap in it (many paintings have overlap in them), but rather because this painting has nothing else in it except overlap. It is only the fact of the overlap, and the resulting multiplicity of aspects which the forms present, that makes the painting fascinating. It seems almost as though the painter had made an explicit attempt, as I have done, to single out overlap as a vital generator of structures.

All the artificial cities I have described have the structure of a tree rather than the semi-lattice structure of the Nicholson painting. Yet it is the painting, and other images like it, which must be our vehicles for thought. And when we wish to be precise, the semi-lattice, being pair of a large branch of modern mathematics, is a powerful way of exploring the structure of these images. It is the semi-lattice we must look for, not the tree.

When we think in terms of trees we are trading the humanity and richness of the living city for the conceptual simplicity which benefits only designers, planners, administrators and developers. Every time a piece of a city is turn out, and a tree made to replace the semi-lattice that was there before, the city takes a further step toward dissociation.

In any organized object, extreme compartmentalization and the dissociation of internal elements are the first signs of coming destruction. In a society, dissociation is anarchy. In a person, dissociation is the mark of schizophrenia and impending suicide. An ominous example of a city-wide dissociation is the separation of retired people from the rest of urban life, caused by the growth of desert cities for the old such as Sun City, Arizona. This separation is possible only under the influence of tree-like thought.

It not only takes from the young the company of those who have lived long, but, worse, causes the same rift inside each individual life. As you will pass into Sun City, and into old age, your ties with your own past will be unacknowledged, loose, and therefore broken. Your youth will no longer be alive in your old age-the two will be dissociated, your own life will be cut in two.

For the human mind, the tree is the easiest vehicle for complex thoughts. But the city is not, cannot, and must not be a tree. The city is a receptacle for life. If the receptacle severs the overlap of the strands of life within it, because it is a tree, it will be like a bowl full of razor blades on edge, ready to cut up whatever entrusted to it. In such a receptacle life will be cut to pieces. If we make cities which are trees, they will cut our life within to pieces.

Words and Phrases

1. visualize   *v.*   想象，形象化
2. simultaneously   *adv.*   同时地
3. primitive   *adj.*   简单的、原始的
4. pigeonhole   *n.*   鸽巢
5. analogy   *n.*   类推、类似
6. mental construct   心智构造
7. category   *n.*   类别
8. intolerance   *n.*   不能容忍
9. chaos   *n.*   /chaotic   *adj.*   混乱/混乱的
10. plastic   *adj.*   造型的
11. fascination   *n.*   迷恋
12. inventory   *n.*   详细目录
13. parallelogram   *n.*   平行四边形
14. explicit   *adj.*   /implicit   *adj.*   外在的、清楚的/内在的、含蓄的
15. generator   *n.*   发生器
16. vehicle for thought   *n.*   传达思维的手段
17. dissociation   *n.*   分裂
18. compartmentalization   *n.*   划分、区分
19. anarchy   *n.*   无政府状态
20. schizophrenia   *n.*   精神分裂症
21. impend   *v.*   即将发生
22. ominous   *adj.*   不吉利的、凶兆的
23. rift   *n.*   裂缝
24. strand   *n.*   线、索

Section 2
Extensive Reading

The Metropolitan Opera House

From its opening in 1883, the Metropolitan Opera has been one of the world's leading opera companies. Today, the Met's preeminent position rests on the elements that established its reputation: high quality performances with many of the world's most renowned artists, a superior company of orchestral and choral musicians, a large repertory of works, and the resources to make performances available to the public.

The first Metropolitan Opera House was built on Broadway and 39th Street by a group of wealthy businessmen who wanted their own opera house. In the company's early years, the management changed course several times, first performing everything in Italian (even Carmen and Lohengrin), then everything in German (even Aida and Faust), before finally settling into a policy of performing most works in their original language—with some notable exceptions.

The Metropolitan Opera House has always engaged many of the world's most important artists. Christine Nilsson and Marcella Sembrich shared leading roles during the opening season. In the German seasons that followed, Lilli Lehmann dominated the Wagnerian repertory and anything else she chose to sing. In the 1890s, Nellie Melba and Emma Calvé shared the spotlight with the De Reszkes (Jean and Edouard), and two American sopranos, Emma Eames and Lillian Nordica. Enrico Caruso arrived in 1903 and by the time of his death had performed more times with the Met than with all other opera companies combined. American singers acquired even greater prominence with Geraldine Farrar and Rosa Ponselle becoming important members of the company. In the 1920s, Lawrence Tibbett became the first of a distinguished line of American baritones for whom the Met was home. Today, the Metropolitan continues to present the best available talent from around the world, and also concentrates on finding and training artists through its National Council Auditions and Lindemann Young Artist Development Program.

Great conductors have helped shape the Metropolitan, from Wagner's disciple Anton Seidl in the 1880s and 1890s to Arturo Toscanini who made his debut in 1908. There were two seasons with both Toscanini and Gustav Mahler on the conducting roster. Later, Artur Bodanzky, Bruno Walter, George Szell, Fritz Reiner, and Dimitri Mitropoulos contributed powerful musical direction. James Levine made his debut in 1971 and has been Music Director since 1976 (holding also the title of Artistic Director between 1986 and 2004).

The Metropolitan Opera has given the American premieres of some of the most important works in the repertory. Among Wagner's masterpieces, Die Meistersinger, Das Rheingold, Siegfried, Götterdämmerung, Tristan and Isolde, and Parsifal were first performed in this country by the Met. Other American premieres have included Boris Godunov, Der Rosenkavalier, Turandot, Simon Boccanegra, and Arabella. The Met's twenty-nine world premieres include Puccini's La Fanciulla del West and Il Trittico, and three recent works, John Corigliano and William Hoffman's The Ghosts of Versailles in 1991, Philip Glass's The Voyage in 1992, and John Harbison's The Great Gatsby in 1999. The Met has commissioned new works from Tobias Picker and Tan Dun for future seasons. An additional thirty-four operas have had their Met premieres since 1976.

Hansel und Gretel was the first complete opera broadcast from the Metropolitan Opera on Christmas Day 1931. Regular Saturday afternoon live radio broadcasts from December to April quickly made the Metropolitan Opera a permanent presence in communities throughout the United States and Canada. The radio broadcasts are now heard not only across North America but in Europe, Australia, New Zealand, South America, China and Japan.

In 1977, the Metropolitan began a regular series of televised productions with a performance of La Bohème viewed by more than four million people. "The Metropolitan Opera Presents" has made seventy-eight complete Met performances available to a huge audience around the world. Many of these performances have been issued on videotape, laserdisc and DVD.

Almost from the beginning, it was clear that the opera house on 39th Street did not have adequate stage facilities. However, it was not until the Metropolitan Opera joined with other New York institutions in forming Lincoln Center for the Performing Arts that a new home became possible. The new Metropolitan Opera House, which opened at Lincoln Center in September of 1966, was equipped with the finest technical facilities.

In 1995, the Metropolitan introduced "Met Titles", a unique system of simultaneous translation. "Met Titles" appear on individual computerized screens mounted in specially built railings at the back of each row of seats, for those members of the audience who wish to utilize them, but with minimum distraction for those who do not. "Met Titles" are provided for all Metropolitan Opera performances.

Each season the Metropolitan stages more than two hundred performances of opera in New York. More than 800,000 people attend the performances in the opera house during the season. Millions more, throughout the world, experience the Metropolitan Opera on television, radio, on tour and recordings.

Resources for Reference

http://www.aviewoncities.com/nyc

Section 3
Tips for Translation

Translating comparative structures 比较结构的翻译

在很多情况下，比较结构，如"less than"、"more than"、"not so much as"等，不能简单地翻译成"比……多（少）"，而应当根据句子的实际含义和汉语习惯，力求通顺、明白。

1. Architects themselves admit more and more freely that they really like living in old buildings *more than* new ones.
建筑师们越来越坦率地承认他们喜欢居住在老房子而不是新建筑里。
　　上句中的"more than"用"比……多"较难翻译出句子的本意。

2. Tugwell's towns ended up as "*little more than* planned middle-class bedroom suburbs."
塔维尔的城镇最终不过就是规划中的中产阶级在郊区的卧室。
　　"little more than"的直译有一定的困难，可以结合"end up as（以……告终）"，翻译成"最后不过就是"，把句子隐含的负面评价的语气体现出来。

3. …while a semi-lattice based on the same 20 elements can contain *more than* 3,000,000 different subsets.
……基于同样20个元素的半网络结构可以包含超过三百万个不同的子集。
　　"more than"在此译为"超过"。

4. It is a large general contractor who involves itself with a lot of important building projects around the world…*more than* an architecture design company.
……这不仅仅是一家建筑设计公司，而是一家大型的总承包公司，在世界各地参与了众多重要建设项目。
　　"more than"在有些情况下还可以译成"不仅仅"。

5. It was *not so much as* the architect he liked as the building he designed.
与其说他喜欢那栋建筑，不如说喜欢设计那栋房子的建筑师。
"not so much as"带有"与其说……不如说……"的意思。

Section 4
Listening Practice

Please watch the video and answer the questions below.
1. What does the café try to remind people of?
2. Why did Zumthor make the construction very simple?

Words and Phrases:
1. stimulate   *v.*   刺激；激励
2. provoke   *v.*   煽动；激起
3. memorial   *adj.*   纪念的，悼念的/*n.* 纪念碑，纪念物
4. compassion   *n.*   怜悯，同情
5. drill   *n.*   钻头
6. theatrical   *adj.*   夸张的，戏剧性的；
7. corrugate   *v.*   （使某物）起皱褶
8. contemporary   *adj.*   当代的，现代的
9. wino   *n.*   酒鬼
10. militant   *adj.*   激进的
11. enterprise   *n.*   企（事）业单位；事业

Resource for Reference

https://www.youtube.com/watch?v=dSfkim0mohA

Unit 10

Section 1
Intensive Reading

Modern Housing Prototypes
Roger Sherwood

Part I

INTRODUCTION This book is presented in the belief that a reexamination of some of the great housing projects of this century is appropriate at a time when the design of housing commands the attention of architects the world around. The buildings offered here as case studies were selected because of their importance as prototypes, projects that set the standards and patterns of much that was, and is, to follow. Other considerations were diversity——so that a wide range of countries, buildings types and problems would be represented——and architectural quality. My assumption is that there is no excuse for poor architecture; that housing, like all buildings, to paraphrase Geoffrey Scott, must be convenient to use, soundly built, and beautiful.

But why prototypes? One of the essential points of heuristic thought——the process of discovery and invention relating to problem solving——is the awareness that, until a problem is clearly defined, guesses or conjectures must be made to help clarify the problem. During the period of uncertainty, Reference to analogous problems can be used to give a new turn to one's thinking. Through the study of solutions to related problems, a fresh conclusion may be reached.

Various writers have suggested that it is never possible to state all the dimensions of a problem, that "truly quantifiable criteria always leave choices for the designer to make". ① In the absence of clear design determinants, and to avoid purely intuitive guessing, it has been argued that analogous reference might give design insight; that perhaps a paradigm of the problem might be accepted as a provisional solution, or an attack on the problem might be made by adapting the solution to a previous problem; that during the period when many of the variable are unknown, a "typology of forms" might be used as a simulative technique to clarify the problem.

The notion of using an analogous problem as a paradigm for gaining insight into a present problem is not, of course, new. A mathematician typically looks for an auxillary theorem having the same or a similar conclusion. ② In architecture, invention often passes through a phase of groping, where ideas about a projected building form are triggered by exposure to some existing building with a similar program, functional specification, or site condition. The analogous building then becomes in some sense a model or a prototype.

The use of prototypes is especially useful in the design of housing because housing lends itself to systematic typological study. Most building types, such as theaters, schools, factories, or even office buildings, have to respond to different programs and are rarely consistent and repetitive. Housing, because it consists of repeating units with a consistent relation to vertical and horizontal circulation, can more logically be studied in terms of its typological variations. Although housing would seem to embrace almost unlimited possible variations, in fact there are not many basic organizational possibilities and each housing type can be categorized easily.

While building regulations, construction techniques, and housing needs have considerable impact on the form that housing may take at any given time in any given culture, still only a few dwelling unit types are plausible, and these units may be collected together in only a few rather limited ways that do not change very much from country to country. An apartment building unit today in Zagreb——as an organization of building units——is much like an apartment building in Berlin or Tokyo. Even extreme cultural requirements, such as the provision for a tatami life-style in Maekawa's Harumi slab in Tokyo (Figure 10-1), have resulted in an organization that can easily

---

① Alan Colquhoun, *Typology and Design Method*, Arena, 1967, pp. 11-14 Karl Popper has perhaps best articulated the notion that logical heuristic process can be stimulated in situations characterized by a lack of quantifiable data by offering tentative solutions and then criticizing these solutions. Popper's book *Conjectures and Refutations* (New York: Basic books, 1962) is a lengthy justification of this procedure. William Bartley, "How Is the House of Science Built", *Architectural Association* Journal, February 1965, pp. 213-218, summarizes Popper's thesis as follows: "the first job of the man who has a problem must be to become better acquainted with it. The way to do this is by producing an inadequate solution to the problem——a speculation——and by criticizing this. To understand a problem means, in effect, to understand its difficulties; and this cannot be done until we see why the more obvious solutions do not work. Even in those cases where no satisfactory answer turns up we may learn something from this procedure" (p. 216). Max Black also deals with the idea of analogous reference or model in *Models and Metaphors* (Ithaca: Cornell University Press, 1962), especially chapter13, "Models and Archetypes".

② See G. Polya, *How to solve it* (New York: Doubleday, 1957).

be compared to a Western model; Park Hill in Sheffield of the sixties (Figure 10-2), For example, is organizationally similar. Both have larger and smaller units in the typical section. Entrance to the larger of the two——a two-level unit——is at the corridor level, with rooms above; stairs lead to the smaller unit below. In each, therefore, the corridor occurs at every other level, and stairs lead up and down from there. Although the position of the stairs, kitchen, and both are different——along parallel walls in Harumi and in a zone parallel to the corridor in Park Hill——and the sitings of the buildings are quite different, nevertheless they are organized fundamentally alike. Even the Arab housing designed in Morocco in the fifties by ATBAT (Figure10-3), where cultural requirements dictated absolute visual privacy, outdoor cooking, and a lack of the usual room subdivisions and conventional toilets, resulted in a building which, although it has a peculiar checkerboard elevation, is more or less a conventional single-loaded, gallery access apartment building.

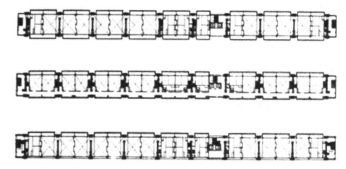

Figure 10-1 Harumi Apartment House, Tokyo. Kunio Maekawa, 1958.

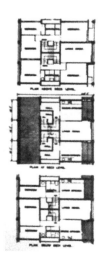

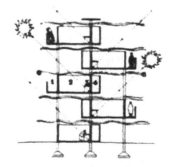

Figure 10-2 Park Hill, Sheffield. Lewis Womersley, 1959.

Figure 10-3 Housing, Morocco. ATBAT, 1950.

Whatever his cultural, economic and technical constraints, every architect is confronted with choices and questions about organization. How will the individual apartments be arranged? How will the mix of different apartment types be accommodated? What circulation systems——horizontal and vertical——can service this mix of apartments? What is the best circulation system? Walk-up or single-loaded, double-loaded, or skip-stop corridor system? Where is entrance and access to the vertical circulation system? What building form dose this collection of units take: low-rise or high-rise, rowhouse, slab or tower? These fundamental organization questions are pertinent to any housing project. Modern Housing Prototypes is intended to provide the architect with a set of analogues references to help him solve these basic organizational problems.

Words and Phrases

1. prototypes   *n.*   原型
2. heuristic   *adj.*   启发式的
3. criteria   *n.*   标准
4. intuitive   *adj.*   直觉的
5. paradigm   *n.*   范例
6. typology   *n.*   类型学
7. archetype   *n.*   原型，原始模型
8. typological   *adj.*   类型（学）的
9. circulation   *n.*   流线
10. plausible   *adj.*   似是而非的
11. siting   *n.*   选址
12. single-loaded   *adj.*   外廊式的
13. gallery access   *adj.*   走廊进入式的
14. walk-up   *n.*   无电梯的公寓，*adj.* 无电梯的
15. double-loaded   *adj.*   内廊式的
16. skip-stop corridor   隔层设置的走廊
17. low-rise   *adj.*   低层的
18. high-rise   *adj.*   高层的
19. rowhouse   *adj.*   联排式住宅
20. slab   *n.*   板式建筑
21. tower   *n.*   塔式建筑

Section 2
Extensive Reading

Editorial Reviews

Journal of Architectural Education: By including idiosyncratic works, Sherwood has encompassed the best that "modern" architecture has offered us to live in. His list of examples is difficult to fault and his drawings and informative text make this book well worth owning.

The Yale Graduate Professional: Presented so clearly and concisely, Sherwood's projects may be easily compared and are comprehensible to those unfamiliar with housing design. The book itself is attractively organized, harmoniously balancing text, photograph, and drawing. The juxtaposition of these, as well as the use of color in selected axonometrics, adds not only to the beauty of this collection but also to the accessibility and clarity of the projects presented within.

Book Description

The design of housing has commanded the attention of the greatest architects of the twentieth century. In this stunning volume, Roger Sherwood presents thirty-two notable examples of multi-family housing from many countries and four continents, selected for their importance as prototypes. Designed by such masters as Frank Lloyd Wright, Le Corbusier, Mies van der Rohe, and Alvar Aalto, they range from single-house clusters to row-houses, terrace houses, party-wall and large-courtyard housing, to urban high-rise towers and slabs.

The thirty-two buildings or housing complexes are illustrated with photographs, site plans, floor plans, elevations, and marvelous axonometric drawings. In each case Mr. Sherwood gives background information on the project, mention, factors the architect had to take into consideration (social, environmental, financial), points out creative solutions to particular problems, and comments on special features of the design. Laymen as well as professionals will find his presentations enlightening.

In the Introduction, Mr. Sherwood sets forth the basic principles of organization that apply to housing. He analyzes first the limited number of ways in which individual apartments or living units can be laid out (each type or plan lending itself to variations and permutations) and then the ways in which different units can be vertically and horizontally organized within a single building. Drawings and plans of more than eighty housing complexes in twenty countries accompany his analysis.

Mr. Sherwood offers his book in the belief that there is no excuse for shoddy architecture; that no branch of architecture is more important than the design of human habitations; and that much is to be learned from the study of significant buildings of the recent past.

Resources for Reference

http://www.hup.harvard.edu/catalog/SHEMOD.html

http://www.amazon.com

Section 3
Tips for Translation

Translating the passive voice 被动语态的翻译

相对于汉语，英语中被动语态用得较多。因此被动语态的翻译是英译中的过程中必须注意的问题。

1. The first Metropolitan Opera House *was built* on Broadway and 39th Street by a group of wealthy businessmen who wanted their own opera house.
大都会歌剧院最早是由一帮想拥有自己歌剧院的有钱的生意人在百老汇和第39大街兴建的。
　　此句的翻译中，英语中的被动语句式被翻译成汉语中的主动句式。

2. The Global Financial Center, which will the highest skyscrapers in China, *is expected* to be finished in 2008.
将成为中国第一摩天楼的环球金融中心可望在2008年完工。
　　此句的翻译中，使用汉语中的"主语—谓语"被动句式来替代英语的被动句。译文中的"可望"在汉语中就带有被动含义。下一句也是类似的情况。

3. The structural difference between semi-lattice and tree-like cities can *be expressed* in a single word: overlapping.
半网络和树形城市的结构区别可以概括成一个词：交叠。

4. The problem of historical preservation can *be solved* in diverse ways.
历史保护的问题可以通过不同的方式得到解答。
　　在被动语态的翻译中，也可以通过使用包含被动意义词，如"使"、"由"、"得到"、"给"等等，来体现英文的被动含义。上句中的"be solved"的翻译即是如此。下一句也是类似的情况。

5. More than 30% the concrete in the world *is manufactured* by China annually.
世界上每年30%以上的混凝土是由中国生产的。

6. *It is said that* Ando Tado is going to pay a visit to our school during his stay in Shanghai.
据说安藤忠雄在上海停留期间将访问我们学院。
　　英文中的一些常用被动句式如"It is said that…"、"It is reported that…"、"It is well known that…"、"It must be pointed out that…"等可以不加主语进行翻译，如"据说……"、"据报道……"、"众所周知……"、"必须指出的是……"。

Unit 11

Section 1
Intensive Reading

Modern Housing Prototypes
Roger Sherwood

Part II

Unit Types

Beginning with basic apartments or units, only two are suitable for repetitive use; one other——the 90° double-orientation unit——has limited application. The basic types are (Figure 11 – 1):

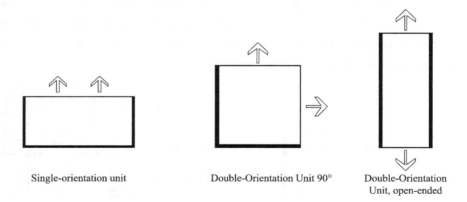

Figure 11 – 1 Unit Typcs

Each of these three unit type has several typical variations, depending upon the positioning of core elements——kitchen, bath and stairs (when used inside the unit)——the entrance options, and the depths necessary for natural light. Minimum unit dimensions vary from country to country as building regulations and construction practices differ, and the arrangement of core elements, natural light, and ventilation requirements change from place to place.

Single-orientation unit

Units that open or face to one side come in two types: with core elements arranged along transverse walls, perpendicular to the corridor. Although these units have a preferred side——they face outward and are most often used where three sides are closed except for the entrance from the corridor (a typical double-loaded corridor arrangement)——some single-loaded, open gallery-access versions may have some minor windows opening to the gallery.

Single-orientation unit: transverse core. This type has the advantage of using the transverse structural wall for core elements, so that most plumbing and mechanical stacks are adjacent to structural walls in a back-to-back arrangement between units. The obvious disadvantage with the type is that the kitchen and in some cases the bath are taking up exterior surface which could be better used for living and sleeping areas, since under many building codes the kitchen and bath do not require natural light and ventilation. An awkward plan can result when the kitchen is on one transverse wall and the bath on the other. Also, the blank exterior walls that core elements tend to create (especially with the small windows typically used in a kitchen or bath) generate elevational problems: these blank surfaces also contradict the preferred side characteristics of the type.

The typical unit may include a scheme where the kitchen and bath are together on one wall with the kitchen to the outside, like the Sorgenfri block in Malmo, Sweden, by Jaenecke and Samuelson (Figure 11-3). Other variations include two-story units such as Lincoln Estate slab by J. I. Martin (Figure 11-4). Here two units interlock around an interior core of stairs and toilets; the kitchen in each unit is in a zone along the transverse wall on one side of the building. Park Hill (Figure 11-2) has a similar arrangement although it employs an alternate level corridor; the floor above and below the corridor level are double-orientation unit types (open both front and rear), with the kitchens lining up on one side of the building.

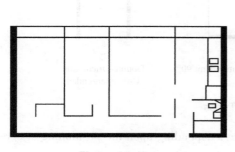

Figure 11-2

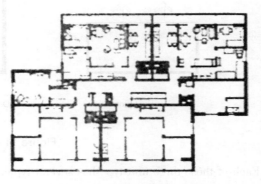

Figure 11-3 Sorgenfri apartment block, Malmo, Sweden, Jaenecke and Samuelson, 1959

Words and Phrases

1. Single-orientation unit   单一朝向的单元
2. Double-Orientation Unit 90°   转角单元
3. Double-Orientation Unit, open-ended   双向开敞的单元
4. ventilation   *n.*   通风
5. perpendicular   *adj.*   垂直的
6. transverse   *adj.*   横的
7. core   *n.*   核,核心
8. plumbing   *n.*   管道工程
9. stack   *n.* 堆, *v.* 堆叠
10. back-to-back   背靠背的
11. take up   占据
12. building codes   建筑规范
13. scheme   *n.*   配置,计划
14. interlock   *vi.*   结合,连结,互锁, *vt.*   使联锁,使连结
15. rear   *n.*   后面,后方, *adj.* 后面的

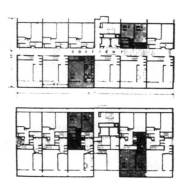

Figure 11-4   Lincoln Estate, London. Martin, Bennett, and Lewis, 1960.

Section 2
Extensive Reading

High Life

In 1909, Life magazine published a cartoon portraying a sleek, 84 storey steel structure open to the elements. On each airy floor of this utopian skyscraper in Manhattan stands a villa surrounded by a country garden and open land. Flocks of birds and double-decker aeroplanes fly right through the building; cottonwool clouds waft over the floors, modestly veiling the spartan steel beams; fountains splash between cypress trees; and an elevator transports the inhabitants up and down.

For the upper echelons of Manhattan at that time, skyscraper living was not exactly the most desirable mode of existence. A Palladian country house or an Art Nouveau villa was vastly more preferable. The 84 storey subsistence machine in the Life cartoon offered an extravagant solution to conflicting desires to flee the city yet remain in touch with its energy, and united both in a paradoxical image. The stacked villas were years ahead of Le Corbusier's idea of the Ville Contemporaine (1922), upon whose roofs flourished green lawns and abundant verdure.

Ninety-one years on, the Dutch pavilion at the Expo 2000 in Hanover has made the vision of the Life cartoonist a reality. The idea that nature can be manufactured out of thin air on any given floor of a skyscraper——seen in the bonsai cultivations inhabiting Norman Foster's grey, high-tech sky lobbies in Frankfurt, for example——has so far achieved only a laughable degree of success. MVRDV's Dutch pavilion, on the other hand, suggests the idea of a second, artificially generated nature, that could, potentially, be endlessly reproduced.

Perhaps earthly happiness is not to be found on the earth, but instead somewhere high in the stratosphere. At any rate, Winy Maas, Jacob van Rijs, and Nathalie de Vries (MVRDV) have come to this conclusion in the face of Holland's extreme population density. What MVRDV have built in Hanover is not architecture, but a concentrated landscape in a six-pack: a world standing on its head. The foundations are a dunescape; 13 massive oak trunks carry the weight of several floors; and, on the roof, wind turbines extend out of a lake and into the air.

This organic layer cake knows nothing of facades: all levels are open to the weather. Take the elevator to the top, go past the wind-energy park and the pond filled with reeds, and you will see collected rainwater running off the roof and arcing out into an eight-metre high curtain flowing over the entire pavilion. If you take the outside stairs instead of the lift, the wall of water is right beside you——if you're unlucky, you will get drenched by a gust of wind. On the next floor down, you're suddenly floating in a forest of leaves, twelve metres high. The "roots" on the floor below make the trees appear to be standing in several enormous geranium pots, but actually the hanging root buckets are empty. Finally, there are geranium and tulip fields, and the "dune" floor, which is, of course, not made of sand, but sprayed concrete.

All in all, the entire structure is nothing more than an abstraction: the manifestation of an idea that is still under construction; a daring laboratory experiment, still in the development stage, which has been dragged prematurely into the public eye. In this sense, the MVRDV creation distinguishes itself from the perfect boredom of the countless presentation videos which adorn most of the Expo's national pavilions and theme park halls. The Dutch pavilion is an astonishing exception in this over-ambitious exhibition, which hit town as an oily mixture of tourist mart, high-tech flea market, and open-air worship. But in the midst of all this is an idea that could go in unexpected directions.

The pavilion was dreamt up in Rotterdam, where MVRDV work in a rather shabby glass box, a former trade office located on the largest interior harbour in the world, surrounded by stacks of millions of containers. Here, they pile up not just miniature landscapes, but also their visionary models. One cardboard model has deep ravines running under the streets of Amsterdam; the canals, in contrast, appear to be ridiculously flat. It looks rather like something out of Fritz Lang's Metropolis (1926): a new Amsterdam with living spaces below sea level——a proposal from MVRDV, commissioned by the Kingdom of Holland. Equally utopian is the proposed construction above Rotterdam: a gigantic superstructure made of motorways, train lines, and scaffolding; a radical, metaphoric intensification of the port city that restlessly rebuilt itself after the war. MVRDV call their visions of the city "Lite Urbanism".

The projects constructed by the Rotterdam trio appear hardly less daring than their theoretical models. The headquarters of the private Dutch radio station VPRO in Hilversum, for example, is a winding, endless succession of stairs, ramps, bridges, patios, halls, and several floors. It unites the elegance of Le Corbusier's white Modernist incunabulum, Villa Savoye (1929), with the power of Piranesi's architectural fantasies and the carefree nonchalance of Rem Koolhaas. Koolhaas designs spatial continua (incorporating entire cities, such as the artificial metropolis Eurolille, 1994) and has no interest in details. Ultimately, his work is about the big picture. The VPRO building doesn't have any walls——who cares? Glass skins of various colours suffice as ersatz facades. Light shines through into the depths of the compact block, and from every location you can see the guts of the entire, wonderful, spatial construction.

A year ago, MVRDV published the plans for Datatown. It could be seen as a conceptual blueprint for the Dutch pavilion. What would happen, MVRDV asked, if the contemporary city was no longer viewed from a topographical perspective, but instead as a collection of data? All the monuments, the unique architectural creations, and individual points of view would no longer count; all that would matter would be pure information. Accordingly, they portrayed the city as if it were made of political campaign graphics on a background resembling graph paper.

With only 60 million square metres of inhabitable surface available on the globe, the architects concluded that we need to erect gigantic stacking cities in order to survive in the long-term. One of these megalopolises would shelter the entire population of the USA, requiring a city park a million times bigger than Central Park, and, for reasons of space, these green lungs would have to be stacked up over 3884 floors. Over 200 years, Datatown's mountain of garbage would grow to a veritable Mont Blanc.

In the light of Datatown, the Dutch pavilion seems to be a practical model for the reinvention of the world; a utopian formula born of necessity to allow the unlimited creation of new real estate. It might be concluded that the city planners of the future will only create endlessly reproducible platforms, which could be fashioned after each individual's own whims: as a dune park, a leafy forest, or a swimming lake with its own power station. This stacked megalopolis in Minimalist dress would not be exactly what the Life cartoonist imagined in the last century——he still dreamed of bucolic rural scenes and mondaine celebrations with the in-crowd on the 84th floor. But has anyone ever honestly believed that Utopia would be fun to live in?

Resources for Reference

http://www.frieze.com/issue/article/high_life

http://www.nytimes.com/2008/06/08/magazine/08mvrdv-t.html?pagewanted=1&fta=y

Section 3
Tips for Translation

Translating attributive clause 定语从句的翻译

定语从句的翻译总体上可分为限制性定语从句和非限制性定语从句的翻译。相比较而言，限制性定语从句与先行词的关系更为紧密，而非限制性定语从句通常是对先行词进行补充说明。

1. Through the use of flat and folded plates, curvilinear forms, vaults, and interlocking building components, he provides a narrative *that creates an architectural dialogue through the medium of a structural language*.
通过平板、折板、曲面、拱顶以及相互关联的建筑构架的使用，他提出了一种以结构为媒介的进行建筑对话的表述方式。
   本句的翻译是按照定语从句的一般翻译方式，把限制性定语从句译成前置的定语。

2. Even wood, a material *that can span relatively short horizontal distances by beam action*, has to be combined in conical, cylindrical or spherical shapes whenever large distances are to be spanned.
即便是木头这种能够通过梁的方式获得相对较短的水平跨度的材料，也必须通过组合成锥状、柱状或者球状才能跨越较大的距离。
   这一句的翻译也是限制性定语从句的一般译法。

3. Pier Luigi Nervi was born in Sondrio and attended the Civil Engineering School of Bologna, *from which he graduated in* 1913.
皮埃尔·鲁齐·纳尔维出生于桑德里奥，就读于博洛尼亚市政工程学院，并于1913年毕业。
   本句的翻译是按照定语从句的一般翻译方式，把非限制性定语从句译成并列的分句并后置。

4. Aside from being a brilliant mathematician and an outstanding engineer *whose office was considered an ideal training ground for young engineers*, Mario Salvadori was also a charismatic teacher of structures at the School of Architecture, Planning and Preservation at Columbia University.
马里奥·萨尔瓦多里不只是才华横溢的数学家和杰出的工程师，他的办公室被认为是初出茅庐的工程师经受锻炼的理想之处。他也是哥伦比亚大学建筑规划与历史保护学院一位充满魅力的结构学教师。

这一句中的"whose office was considered an ideal training ground for young engineers"是限制性定语从句，但也译成后置的并列分句并，主要是考虑到如果硬是把这一成分置于现行词前面，会使译文头重脚轻，不符合汉语的句子结构习惯。

5. The old building, *which was built in the late Qing Dynasty*, is listed as one of the historical monuments by the Shanghai municipal government.
这栋建于晚清时期的建筑被上海市政府列为历史保护建筑之一。

与上一句相反，本句中的"which was built in the late Qing Dynasty"从结构上看虽然是非限制性定语从句，但可以翻译成前置定语，使译文简洁、流畅。

6. The students should pay enough attention to the technical dimension of architecture, *which will play a very important role in their future professional career*.
学生们应当对建筑学的技术层面予以足够的重视，因为这对他们未来的职业生涯至关重要。

某些定语从句从整句的意义上看带有原因、目的和结果等关系，相当于状语从句的作用。在翻译的时候可以加上连接词体现句子中包含的上述逻辑关系。

Unit 12

Section 1
Intensive Reading

Modern Housing Prototypes
Roger Sherwood

Part III

Single-orientation unit; interior core along the corridor. In the more common type of single-orientation unit, the core elements are arranged in a zone parallel and adjacent to the corridor. Entrance is through this zone into the main spaces of the apartment, thus letting the major rooms open to the preferred side of the building. The kitchen and bath are interior spaces with mechanical ventilation. This simpler plan usually features

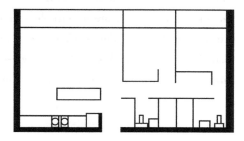

Figure 12-1

a compact back-to-back kitchen and bath grouping and clear, consistent zoning of spaces. The double-loaded corridor slabs designed by Mies van der Rohe (Figure 12-2) are planned this way. He modifies the idea slightly in the Lake Shore Drive apartments (Figure 12-3), where the bath and kitchen are back-to-back but the kitchen opens to the major spaces on the preferred side. Although more typically a plan for double-loaded corridor buildings (where apartments are located on both sides of the corridor), the type is also used for single-loaded or access-gallery plans. The Lamble Street project by Powell and Moya (Figure 12-4) is an example of this type. Or, for a lower density type, there are the courtyard houses by Korhonen and Laapotti in Finland (Figure 12-5).

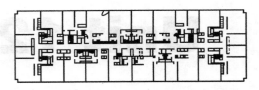

Figure 12-2　Apartments, Baltimore. Mies van der Rohe, c. 1965

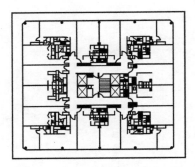

Figure 12-3　Lake Shore Drive apartments, Chicago. Mies van der Rohe, 1948

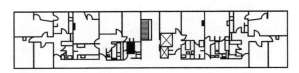

Figure 12-4　Lamble Street housing, London. Powell and Moya, 1954

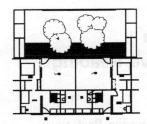

Figure 12-5　Courtyard housing, Espoo, Finland, Korhonen and Laapotti, c. 1968.

Aalto's apartments at Bremen (Figure 12-6), an unusual variation of the single-orientation type, consist of fan-shaped units opening out to the site. Core elements here, although placed along transverse structural walls, are nevertheless in an interior zone along the corridor. The preston housing by Stirling and Gowan (Figure 12-7) is a two-storey version of the same type. These two are single-loaded corridor examples, but the single-orientation unit type is probably most advantageous where three sides of the unit are closed, implying a double-loaded, corridor-every-floor organization.

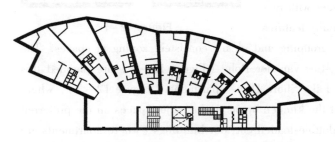

Figure 12-6　Neue Vahr Apartments, Bremen. Alvar Aalto, 1958

Figure 12-7　Preston housing, Lancashire. Stirling and Gowan, 1961

A common variation of the single-oriented unit (applicable to units with either transverse or interior core) works from a strategy of increasing the exterior surface on the open side of the unit so that more rooms can get light and air. Le Corbusier's Immeuble Villas projects of the twenties (Figure 12–8) were of this type: L-shaped units around an open terrace. Although the Immeuble Villas are two-storey units with minor windows on the corridor side of the upper floor of each unit, implying a double orientation, the zone of large volumes and terraces to one side contribute to a definite preferred condition. This type can work in a single-loaded or double-loaded situation. Bishopsfield and Charters Cross housing at Harlow by Michael Neyland (Figure 12–9) is another example of a repeating L-plan, in this case, double-loaded with corridor walls containing only minor windows to the kitchens.

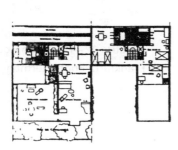

Figure 12–8   Immeuble Villas project, Le Corbusier, 1922.

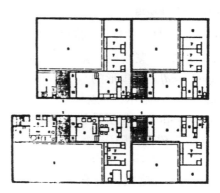

Figure 12–9   Bishopsfield and Charters Cross, Harlow, Essex. Michael Neyland, 1960.

Aalto's Hansa apartments in Berlin (Figure 12–10) are basically a single-oriented type that follows the strategy of increasing exterior surface: its U-plan features dining, living, and bedrooms all around a central terrace. Schindler's EL Pueblo Ribera houses at La Jolla (Figure 12–11) are also single-orientation, U-shaped units coupled together in pairs with hedges used to define and enclose the courtyard spaces.

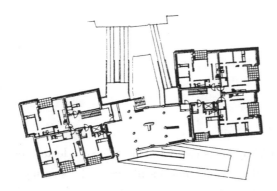

Figure 12–10   Hansaviertel apartments, Alvar Aalto, 1956.

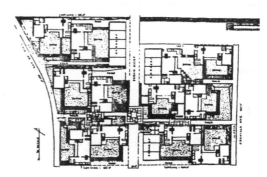

Figure 12–11   El Pueblo Ribera piano houses, La Jolla. R. M. Schindler, 1923.

A possible variation of the single-orientation type is the matte housing scheme, where a matrix of walls is built with each unit inside a walled-in area. Access requirements limit the number of collective arrangements possible, but Egon Eiermann's matte housing in Frankfurt of 1966 (Figure 12-12) is an example of this type. Although there are small private gardens on the entrance side of each apartment, most major spaces open to a private garden to the rear, establishing the single orientation.

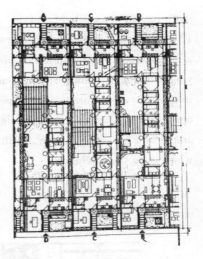

Figure 12-12　Patio housing, Frankfurt. Egon Eiermann, 1966.

Words and Phrases

1. feature　*vt.*　是…的特色，特写，放映，*vi.* 起重要作用
2. compact　*adj.*　紧凑的，紧密的，简洁的
3. fan-shaped　*adj.*　扇形的
4. implying　*vt.*　暗示，意味
5. corridor-every-floor　每层设置走廊的
6. hedge　*n.*　树篱，障碍物
7. enclose　*vt.*　围绕，围合
8. courtyard　*n.*　庭院，院子
9. strategy　*n.*　策略
10. matte　*adj.*　不光滑的，表面粗糙的
11. matrix　*n.*　矩阵
12. walled　*adj.*　有墙壁的
13. collective　*adj.*　集合的，集体的

Section 2
Extensive Reading

Crowded House

Part I

In the fall of 2002, a young Dutch architect named Winy Maas came to Yale to give a lecture on designing and building the 21st-century city, the challenges of which he illustrated by showing a 30-second video that could have been shot above any American metropolitan airport: a view of the tops of several buildings and then, as the camera rose, more and more buildings, more roads and bridges and asphalt lots, until an ugly concrete skin of low-rise development spread to all horizons. Maas was not the first architect to protest the unsightly sprawl that humans have left over much of the earth's surface, but he may have been the first to suggest that we preserve what's left of our finite planetary space by creating "vertical suburbias" ——stacking all those quarter-acre plots into high-rise residential towers, each with its own hanging, cantilevered yard. "Imagine: It's Saturday afternoon, and all the barbecues are running", Maas said, unveiling his design for a 15-story building decked out with leafy, gravity-defying platforms. "You can just reach out and give your upstairs neighbor a beer. " He turned next to agriculture. Noting that the Dutch pork industry consumes huge swaths of land——Holland has as many pigs as people——Maas proposed freeing up the countryside by erecting sustainable 40-story tower blocks for the pigs. "Look—— it's a pork port, " he said, flashing images from PigCity, his plan for piling up the country's porcine population and its slaughterhouses into sod-layered, manure-powered skyscrapers that would line the Dutch coast.

157 VARIETIES: Silodam, a Lego-like residence with 157 apartments of different sizes and prices, rises on stilts from the harbor in Amsterdam (Figure 12 – 13).

STACKED: At Frosilos, a residence in Copenhagen, the apartments were built on the outside of two silos; the continuous glass facades allow maximum light (Figure 12 – 14).

Figure 12 – 13   Figure 12 – 14

Maas is the charismatic frontman for the Rotterdam-based architecture, urban-planning and landscape-design firm known as MVRDV, which brims with schemes for generating space in our overcrowded world. With his messy, teen-idol hair and untucked shirt, Maas strolled the stage extolling the MVRDV credo——maximize urban density, construct artificial natures, let data-crunching computers do the design work——while various mind-bending simulations played across the screen: skyscrapers that tilted and "kissed" on the 30th floor; highways that ran through lobbies and converted into "urban beaches"; all the housing, retail and industry for a theoretical city of one million inhabitants digitally compressed into the space of a three-mile-high cube.

The Netherlands, prosperous and progressive, has long been one of the world's leading exporters of architectural talent. By the mid-1990s, not only Rem Koolhaas and his Office for Metropolitan Architecture but also a whole new generation of designers——MVRDV, West 8, UNStudio——were trying to enlarge Le Corbusier's definition of architecture as the "magnificent play of volumes brought together under light" and arguing for a process driven by research, information and a greater social and environmental awareness. Fighting their battles not just building to building but on a sweeping, citywide scale, Holland's architects and designers were, in the words of the Dutch culture minister, "heroes of a new age."

Still, paradigms tend to fall only under pressure, and at the start of the new millennium an audience at the Yale School of Architecture could be forgiven for greeting vertical suburbs, pig cities and the rest of MVRDV's computer-generated showmanship with the same slack-jawed disbelief that once greeted Fritz Lang's "Metropolis" or the 1909 Life magazine cartoon that promised an urban utopia of country villas perched atop Manhattan skyscrapers while double-decker airplanes whizzed through their atria. When Maas came to New Haven, MVRDV was barely 10 years old and had hardly built outside its native Holland. And yet there he was with his straight-faced scheme to "extend the globe with a series of new moons"——send up food-producing satellites that would orbit the earth three times a day. "Can you imagine," he said with a boyish, science-fair enthusiasm that indulged no irony, "if we grew our tomatoes 10 kilometers high?" On the lecture-hall screen, New York's skyline appeared just as the MVRDV satellite passed overhead, darkening Gotham with a momentary eclipse of the sun.

Who were these Dutch upstarts? And in the so-called real world, would anything actually become of their grand, improbable visions?

The 45 architects and designers who make up MVRDV (the name is formed by the surname initials of Mass and his two founding partners, Jacob van Rijs and Nathalie de Vries) work out of a converted, loftlike space in an old printing plant in Rotterdam, a dull but industrious port city whose historic districts were leveled by the Nazis and whose jagged skyline of new office towers and construction cranes attests to its still-restless effort to rebuild. Inside MVRDV, a liquid northern light pours through a wall of high arched windows, and the occasional cries of foghorns and seagulls confirm its location just blocks from the city's main shipping lane. But otherwise, the mostly 30-something architects who sit with a slouching intensity at rows of long communal tables, surfing Google Earth or manipulating blue-foam architectural models, seem to have their minds in other places.

Maas and van Rijs, who both worked for Koolhaas, and de Vries, who practiced with the Delft-based Mecanoo, formed MVRDV in 1991 after their design for a Berlin housing project won the prestigious Europan competition for architects under 40. Holland has always been a good place to think creatively about space, with its congested countryside (16 million people squeezed into an area the size of two New Jerseys), its faith in planning and the democratic welfare state and its keen appreciation for land that comes from having reclaimed two-thirds of its own from the edge of the North Sea. Meanwhile, young designers were hoping the economic boom and housing shortage of the 1990s would give them the chance to build domestically on a large scale. Still, two years after they formed MVRDV, Maas, van Rijs and de Vries were struggling to find work and practicing out of makeshift offices (during meetings with prospective clients, they'd sometimes recruit friends to keep the phones ringing and wander through in suits) when a Dutch public broadcasting company, VPRO, approached them about a possible new headquarters in Hilversum.

The project's constraints were formidable. VPRO's 350 employees——"creative types," van Rijs says; "individualistic," de Vries adds; "a settlement of anarchists with an obnoxious attitude toward corporate identity" Maas concludes——were then spread out among several buildings, enjoying their fiefs and the company's culture of noncommunication. Even if a new headquarters could bring them all under one roof, it was impossible to predict how the employees would actually use the building, given their fluid work patterns and chaotic organizational hierarchies. "The mandate was: How can we get them to start communicating with each other?" Maas says. "And the answer was: By putting them in a box."

Villa VPRO, which became the defining project of MVRDV's early career, is a densely constructed, five-story box——a "hungry box," as one critic called it——with an endlessly flowing and adaptable interior that renders in spatial form the company's anarchic spirit. MVRDV created a concrete labyrinth of winding stairs, twisting ramps and narrow bridges; a continuous surface of stepped and slanted planes with no real walls, just colored-glass partitions so that sunlight could penetrate into the depths of its compact terrain. "Clearly, VPRO was a social-engineering project," Maas says. "We built a vertical battlefield for the users, one place where they could all meet and argue and find out how to behave. Because of all the hills, slabs and stairs, they were forced to maneuver through the building. Some people hated it——they lost their way, they were overwhelmed by their colleagues. Others loved it. But they all had to deal with each other. I like that. That's part of life."

A year later, MVRDV took social engineering to a new level when it won a commission to represent Holland in Expo 2000 in Hanover, Germany. Expos are notorious excuses for creating second-rate architecture, piling up dreary national pavilions and Disneyfied theme parks around which crowds circulate in a candy-consuming stupor. At the Hanover expo, MVRDV stole the show with another vertical confection——this time a six-story tower of stacked and sustainable artificial

Dutch landscapes that included an oak forest, a meadow of potted flowers, ersatz concrete sand dunes for purifying irrigation water and a "polder landscape" of dyke-protected turf powered by wind turbines spinning away on the roof.

The MVRDV pavilion was, one critic wrote, "science class with the chutzpah of Coney Island." Another predicted that it would "go down as one of the few truly great pieces of expo architecture," alongside Mies van der Rohe's Barcelona Pavilion and Moshe Safdie's Habitat flats at the Montreal expo. Visitors lined up for hours to climb through what was inevitably dubbed the "Dutch Big Mac". But beyond its playful innovation, MVRDV had lofty aspirations for its pavilion, hoping that it would carry the optimistic (and very Dutch) message that in the face of extreme population densities and the craving for open land, you could actually manufacture space——even create an artificial nature out of thin air——by condensing your landscapes on the floors of a building and reproducing them endlessly toward the sky.

"The Dutch population is essentially antiurban," de Vries says. "Therefore as architects in Holland we have a special responsibility to make living in cities and under dense circumstances not just habitable but preferable."

"It was sort of a test case," Maas says. "At a time when urbanism is still dominated by 'zoning,' which is a very two-dimensional approach, we wanted to know: can we extend our surfaces? Can we develop an urbanism that enters the third dimension?"

The Hanover pavilion was "a utopian formula born of necessity to allow the unlimited creation of new real estate," wrote the critic Holger Liebs. It was "a practical model for the reinvention of the world".

At the architectural library at the Delft University of Technology, there's a copy of a 736-page book by MVRDV called "Farmax: Excursions on Density", which is a hodgepodge of essays, transcripts, photos, computer designs, graphs and charts, all examining the growing suburban "grayness" of the Dutch landscape and proposing different solutions for saving the pastoral landscape by "carrying density to extremes". So many students have borrowed, read and plundered that copy of "Farmax" that it had to be pulled from circulation and has sat in a state of complete disintegration inside a kind of glass vitrine. When I mentioned this to van Rijs, he laughed and said: "Yeah, I've seen that. Our book is like a museum piece. Isn't that fun?"

While projects like VPRO and the Hanover pavilion were leading to design commissions in Copenhagen, Madrid, Paris, Tokyo and China's Sichuan province, MVRDV was also reaching outside the realm of established architectural practice by producing a series of theoretical exercises——books, films, exhibitions, even computer games——that amounted to an ongoing propaganda war on behalf of the firm's radical ideas about space. After "Farmax", MVRDV put out another doorstop manifesto, "KM3: Excursions on Capacities", which warned that if the global population "behaved with U. S. -citizen-like consumption", another four earths would be required to sustain it. In the exhibit 3D City, they pushed ever upward, advocating giant stacking cities that, as MVRDV breathlessly described them, exist "not only in front, behind or next to you, but also above and below. In short a city in which ground-level zero no longer exists but has dissolved into a multiple and simultaneous presence of levels where the town square is replaced by a void or a bundle of connections; where the street is replaced by simultaneous distribution and divisions of routes and is expanded by elevators, ramps and escalators…"

Perhaps MVRDV's most ambitious theoretical exercise was the traveling computer installation they called MetaCity/Datatown. Predicting that globalism and an exploding planetary population will push certain regions throughout the world into continuous urban fields, or megacities, MVRDV conceived a hypothetical city called Datatown, designed solely from extrapolations of Dutch statistics. ("It is a city that wants to be explored only as information; a city that knows no given topography, no prescribed ideology, no representation, no context. Only huge, pure data.") According to its creators, Datatown was a self-sufficient city with the population of the United States (250 million) crammed into an area the size of Georgia (60,000 square miles), making it the densest place on earth. MVRDV then subjected this urban Frankenstein to 21 scenarios to see how they would affect the built environment: What if all the residents of Datatown wanted to live in detached houses? What if they preferred urban blocks? What could be done with the waste? (Build 561 ski resorts.) What kind of city park would be needed? (A million Central Parks stacked up over 3,884 floors.) "The seas, the oceans (rising as a result of global warming), the polar icecaps, all represent a reduction in the territory available for the megacity. Does that mean that we must colonize the Sahel, the oceans or even the moon to fulfill our need for air and space, to survive? Or can we find an intelligent way to expand the capacities of what already exists?"

On one level, MetaCity/Datatown was a game and a provocation——architecture as a kind of thought experiment: can the urban landscape be reduced to a string of ones and zeroes? Is what we think of as outward reality nothing more than the physical manifestation of information? But MetaCity/Datatown was also a serious investigation: by translating the chaos of the contemporary city into pure information——or, as MVRDV called it, a datascape——and then showing the spatial consequences of that datascape through computer-generated designs, MVRDV set out to reveal how our collective choices and behaviors come to mold our constructed environments. "These datascapes show that architectural design in the traditional sense only plays a very limited role", Bart Lootsma, an architectural historian, writes in one of many essays inspired by the exhibit. "It is the society, in all its complexities and contradictions, that shapes the environment in the most detailed way, producing 'gravity fields' in the apparent chaos of developments, hidden logics that eventually ensure that whole areas acquire their own special characteristics, even at a subconscious level." Lootsma cites a number of these invisible forces——market demands precipitating a "slick" of houses-with-gardens in the Netherlands, political constraints generating "piles" of dwellings on the outskirts of Hong Kong, the cultural preference for white brick causing a "white cancer" of housing estates in the Dutch province of Friesland. These are "the 'scapes' of the data behind it," he writes.

Moreover, to the extent that MVRDV approaches architecture not as a conventional expression of aesthetics, materials and form but as an almost scientific investigation into the social and economic forces that influence our constructions, the datascapes were also a dry run for the firm's own built work. That work, says Aaron Betsky, the former director of the Netherlands Architecture Institute and a longtime MVRDV-watcher, is really an ongoing project of "giving shape to those zeroes and ones," of making the conceptual real, of turning abstract information into concrete form.

Resources for Reference

http://www.mvrdv.nl/_v2/
http://www.mvrdv.nl/_v2/projects/010_villavpro/index.html
http://www.nytimes.com/2005/04/10/arts/design/10ouro.html
http://www.oma.eu/
http://www.nytimes.com/2008/06/08/magazine/08mvrdv-t.html?pagewanted=1&fta=y

Section 3
Tips for Oral Presentation

A great first impression

We form opinions about people the first time we see or hear them. We even form opinions about people we have never met!

People's "perception" about us DOES matter. As a professional speaker who provides workshops, keynotes and consultations on presentation skills and public speaking, I know that we are all judged by people through "What we say", and "How we say it". We are also judged by "How we Dress", "How we walk" and even "How we eat our food". In the work environment, we judge people by the size of their office, the location of their office or by the number of people working for us. As a business owner your company is judged by the way your receptionist answers the telephone or greets people at the door.

Think about it!

You CANNOT, NOT! make a first impression. People always form an initial impression about us the first time they come in contact with us whether it is in person or whether it is over the telephone or even by the way we leave a message on THEIR answering machine. Every other contact with us after that first time either supports or conflicts with that first impression. Create a good first impression and the relationship grows from there. Create a bad first impression and your relationship with that person can be an uphill battle.

Resource for Reference

http://www.selfgrowth.com/articles/laskowski3.html

Unit 13

Section 1
Intensive Reading

Modern Housing Prototypes
Roger Sherwood

Part IV

Double-orientation unit types come in many variations and can be collected together in many different ways. The corner type or 90° double-orientation unit may be seen simply as a singly-oriented in which one of the three closed walls has been opened up. This limits the strategies of collecting units together, since each needs a corner, and the use of this type seems to be limited to towers, smaller freestanding buildings, and to certain kinds of terrace housing.

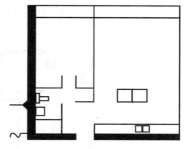

Figure 13–1 Double-Orientation Unit, 90°

Frank Lloyd Wright's Suntop Homes is a good example of this type: four units within crossed party walls, each three stories high opening at the corner. Wright's earlier versions like the Cloverleaf development (Figure 13–2) introduced an internal courtyard. St. Mark's Tower (Figure 13–3) and the built version of it, the Bartlesville Price Tower, adopt the same parti of four corner units with core elements on the interior. Buildings employing this kind of unit necessarily must be freestanding, with private entrance required for projects like Suntop and common lobbies for towers like Bartlesville.

Other examples of one-Story or two-story corner units include the atrium houses at Schwerzenback in Switzerland by Kunz (Figure 13-4) and the Candilis, Josic, and Woods projects, which often consist of buildings planned to gain the corner advantage even to the extent of creating site arrangements consisting of many staggered-plan buildings in an overall system designed to maximize peripheral surface (Figure 13-5). Most compact towers use this type: for example, the Vallingby tower by Ancker and Gate (Figure 13-6) or the Nirwana apartment buildings by Duiker (Figure 13-7), which have a much larger area in plan but are organized with an apartment in the each corner.

Figure 13-2 Cloverleaf project. Frank Lloyd Wright, 1939.

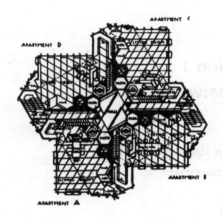

Figure 13-3 St. Mark's Tower, project. Wright, 1929.

Figure 13-4 Atrium houses, Schwerzenback, Switzerland. Fred Kunz, 1967.

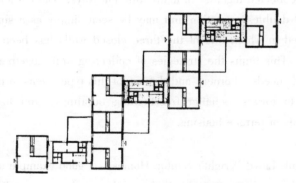

Figure 13-5 Cluster housing project. Candilis, Josic, and Woods, 1959.

Various permutations of the tower use a strategy of creating more exterior surface and hence more corner conditions. While many of these are not strictly 90° units, they are versions of the corner unit in that they cannot be repeated in linear fashion like the singly-oriented types. The Baldessari in the Hansa project in Berlin (Figure 13-8) or the Albany Houses in Brooklyn by Fellheimer,

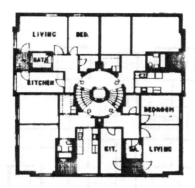

Figure 13-6 Tower, Vallingby, Sweden. Ancker and Gate, 1953.

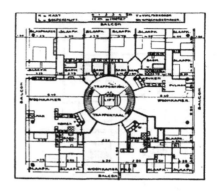

Figure 13-7 Nirwana Apartments, Den Haag. Johannes Duiker, 1927.

Wagner, and Vollmer, done for the New York City Housing Authority (Figure 13-9), are examples of this variation. Pinwheel plans such as the Candilis, Josic, and Woods project at Bagnols sur Ceze of 1957 (Figure 13-10) try to maximize the corner situation. O. M. Ungers employed this idea with a slightly different variation in the Markischesviertel project in Berlin in 1962 (Figure 13-11). Here bedrooms are put into the corners, which are solid except for small windows; the leftover void is designated as living space. Essentially, it is a corner, pinwheel parti that generates——when used in combination——a distinctive staggered site plan (Figure 13-12). This was a popular idea at Markischesviertel, and many architects besides Ungers used it. All these projects are perhaps derived from various Candilis, Josic, and Woods schemes for cluster housing in the mid-fifties, where pinwheel blocks or towers hook up with each other to make a kind of continuous building (Figure 13-13).

Figure 13-8 Hansaviertel tower, Berlin. Luciano Baldessari, 1956.

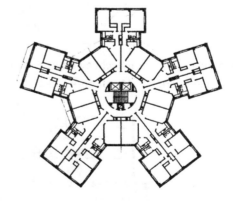

Figure 13-9 The Albany Houses, New York. Fellheimer, Wagner, and Vollmer, 1950.

Still other strategies to increase peripheral surface and multiply corners are the slipped-slab schemes such as this by A/S Dominia in Copenhagen (Figure 13-14). Lasdun in the Bethnal Green towers (Figure 13-15) uses the same idea, as does Aalto with the Hansa block in Berlin.

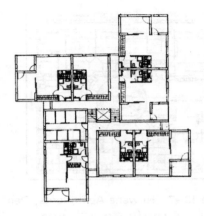

Figure 13-10　Apartment block, Bagnols sur Ceze, France. Candilis, Josic, and Woods, 1957.

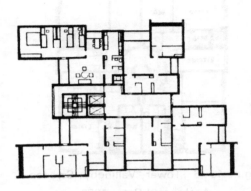

Figure 13-11　Markischesviertel, Berlin, floor plan. O. M. Ungers, 1962.

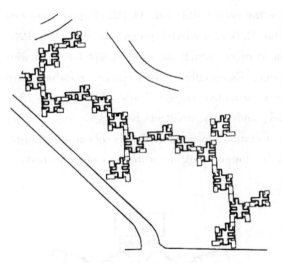

Figure 13-12　Markischesviertel, Berlin, site plan. O. M. Ungers, 1962.

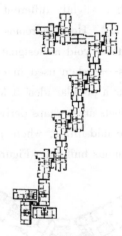

Figure 13-13　Housing block, Clos d'Orville, Nimes. Candilis, Josic, and woods, 1961.

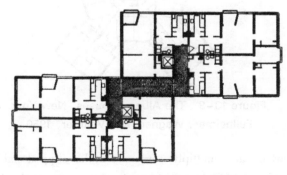

Figure 13-14　Tower, Copenhagen. A/S Dominia, c. 1960.

Figure 13-15　Bethnal Green towers, London. Denys Lasdun, 1960.

Some terrace housing projects utilize a more complex version of 90° or corner unit. The Zollikerberg project in Zurich by Marti and Kast (Figure 13 – 16) is an example of this. Here two-story L-shaped atrium units are placed on top of one anther and stepped up a slope, with a retaining wall against the slope. Side walls are punctured only with small windows. There is a preferred condition toward the garden, but the living room becomes the dominant void at one corner. The Stucky and Meuli units also in Switzerland (Figure 13 – 17) step up a slope, again with the windowless retaining wall against the slope and with essentially closed walls on the two sides. All major rooms open to a continuous terrace, and the living room, which is the main space, opens to two sides at the corner. The above examples are not, strictly speaking, just 90° units because each apartment has openings to three side and does not attach horizontally to other units; however, the positioning of the living room as a large volume at the corner emphasizes the corner condition. The drawing of the Aalto terrace house at Kauttua (Figure 13 – 18) shows this condition three-dimensionally with openings to three sides. But this only suitable on very narrow sites, and a more typical condition perhaps would be side-by-side Kauttuas with each unit more literally a corner type. Denys Lasdun's beautiful apartment block at St. James Place in London of 1960 (Figure 13 – 19) is a

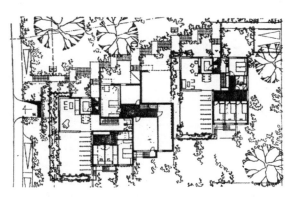

Figure 13-16   Zollikerberg terrace housing, Zurich. Marti and Kast, 1964.

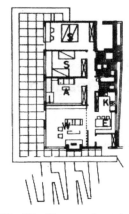

Figure 13-17   Terrace housing, Zug, Switzerland. Stucky and Meuli, c. 1960.

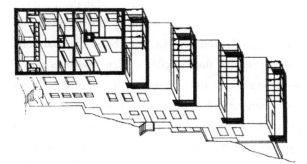

Figure 13-18   Terrace housing, Kauttua, Finland. Alvar Aalto, 1938.

Figure 13-19   Apartment tower, St. James Place, London. Denys Lasdun, 1960.

high-rise example of the same condition. Although it is a tower backed up to an existing party wall on the side open space on the other three sides, it is spatially a 90° type. By use of an ingenious split-level section, Lasdun has been able to further accentuate the corner orientation of the living room, which is one and one-half floors high and opens to a park on the preferred side of the building.

Words and phrases

1. party wall   *n.*   分户墙
2. parti   *n.*   构图
3. atrium   *n.*   中庭
4. staggered-plan   交错变化的平面
5. peripheral   *adj.*   外围的
6. permutation   *n.*   改变，交换
7. Pinwheel   *n.*   风车
8. retaining wall   挡土墙
9. split-level   *adj.*   房内有不同高度平面的，错层的

Section 2
Extensive Reading

Crowded House

Part II

When MVRDV begins a project, it starts by assembling information on all the conceivable factors that could play a role in the site's design and construction——everything from zoning laws, building regulations and technical requirements to client wishes, climatic conditions and the political and legal history of the site. Architects often view these rules and regulations as bureaucratic foils to their creativity. MVRDV sees them as the wellspring of invention. In fact, believing that subjective analysis and "artistic" intuition can no longer resolve the complex design problems posed by the ever-metastisizing global city, the architects sometimes use a home-built software program called Functionmixer. When loaded with all the parameters of a particular construction project, Functionmixer crunches the numbers to show optimal building shapes for any given set of priorities (maximizing sunlight, say, or views, or privacy) and pushes limits to the extreme, where they can be seen, debated and, often, thoroughly undone. It creates a datascape that is the basis of the design.

In 1994, for instance, MVRDV was asked to build housing for the elderly——an apartment block with 100 units——in an already densely developed suburb of Amsterdam. Because of height regulations and the need to provide adequate sunlight for residents, only 87 of the called-for units could fit within the site's restricted footprint. Rather than expand horizontally and consume more of the neighborhood's green space, MVRDV borrowed a page from its "vertical suburbia" and hung the remaining 13 apartments off the side. Their wonderfully odd WoZoCos housing complex takes the conventional vertical housing block and reorganizes it midair with these bulging extensions that seem to be levitating right up off the ground.

Four years later, when MVRDV was selected to build economically mixed housing in Amsterdam's docklands area, the firm held countless negotiations with the parties involved——local politicians, the planning authority, possible future residents——all of whom advocated for a different distribution of the housing. Eventually MVRDV threw all the data into a computer and came up with the Silodam including 157 apartments of various sizes and prices that sit together in one 10-story multi-colored block that rises on stilts from the harbor like a docked container ship. From the outside, the Silodam looks simple enough——as literal as a child's giant Lego construction——but inside the block is filled with a vast array of dwellings arranged into economically mixed "mini-neighborhoods", while a series of communal galleries and gangways allow residents to walk from one end of the "ship" to the other.

MVRDV's radical, research-driven methodology has been a source of fascination to critics and competitors from the start. "No one else has found as convincing a way, " writes the historian Lootsma, of "showing the spatial consequences of the desires of the individual parties involved in a design process, confronting them with each other and opening a debate with society, instead of just fighting for one or the other, as most architects would". And the urbanist and designer Stan Allen, now dean of the Princeton School of Architecture, points out that "rather than impose structure, leading to closure and more precise definition, MVRDV works to keep the schema open as long as possible, so that it can absorb as much information as possible. "

In fact, MVRDV's architects rely so much on gathering and metabolizing data, information and competing points of view that they insist they leave no formal signature on their work. "We try to avoid any sort of aesthetic aspect in our designs, " van Rijs told me. "Unlike Gehry, Zaha and others whose work is easy to recognize, we don't have a strong personal style. Our methodology is based more on logic. Sometimes we call it an iron logic: depending on the situation, we come and take a look and say: 'What's happening? What should be done?' Then we follow a step-by-step narrative, and when you see the building, you get the final result. It's the only possible outcome. You cannot see anything else. "

But if MVRDV's design process is really so rational and objective——if, as Stan Allen says, the architects reject "fuzzy intuition" and "artistic expression" for a step-by-step pragmatism in which "form is explained only in relation to the information it encodes: architecture as a series of switches, circuits or relays activating assemblages of matter and information" ——then why, Allen asks, are their creations so unexpected and witty, sometimes even so spectacular? Commissioned to build large-scale housing in a sprawling Madrid neighborhood already choked with monotonous low-rise construction, MVRDV designed a typical horizontal housing block with an interior courtyard. Then the architects flipped the block on its side to create Mirador, a towering 22-story icon for the neighborhood with the courtyard now transformed into an enormous, open-air balcony offering sweeping views of the Guadarrama Mountains. Some MVRDV designs are so logical they seem to turn reality on its head.

In 2007, two years after Hurricane Katrina devastated much of New Orleans, the actor and architectural enthusiast Brad Pitt asked 14 design firms to help his nonprofit Make It Right rebuild the city's impoverished Lower Ninth Ward, one of the neighborhoods hardest hit by the storm. Specifically, he asked for designs for an affordable——but also floodproof——1,200-square-foot house with three bedrooms and a porch. Maas, van Rijs and de Vries——citizens of a country that is continually defending its buildings from the threat of inundation——had already contributed to an exhibit of post-Katrina architecture: inspired by a child's crayon drawing of New Orleans residents walking to safety up an imaginary hill, they conceived a new elementary school made safe from rising waters by tucking it inside an artificial, grass-covered mound, where balconies hung off the sides and a playground covered the top. Now, having received Brad Pitt's call, they came up with an ingenious, almost whimsical solution to the problem of future flooding: their "Bend House" was a variation on the South's traditional low-slung shotgun houses, this one hinged in the middle so that its front and back are raised above the waterline.

Some critics were appalled. By creating a dwelling that already looked flood-damaged, perhaps even uninhabitable, MVRDV appeared to be using the New Orleans disaster to score political points or, worse, to be winking ironically at the residents' ongoing plight. Others thought the Bend House was emblematic of MVRDV's best work and of the architects' knack for creating buildings whose formal inventiveness arises from the explicit display of the social or environmental problems that brought them to life: VPRO's endless interiors signaling the need for social connection; WoZoCos's hanging boxes showing how to preserve our green spaces; the festively striped Silodam offering ways to mix rich and poor.

"The architecture that we make is part of the ordinary, part of our pop culture," Maas told me. "At the same time, the buildings try to engage with society by questioning our behavior and offering alternatives. And they offer those alternatives by showing visibly, obviously in their actual design the social problems we were trying to address. When you see the object, you see the question."

Maas's remark brought to mind an appraisal of MVRDV's work by the French architect Alain Guiheux. "A great mystery in architectural projects surrounds the definition of what is acceptable to the client," he writes. Where does the client's caution and censorship begin? At what point does that caution become the architect's own self-censorship? Guiheux goes on to say that MVRDV tries to resist society's censorship and overcome its own by using playfulness to "soften up conformity" and by "pushing back the line between the reasonable and the incredible". That, he says, is their "magic", and has effected a break with architectural convention "like that undergone by painting at the beginning of the 20th century, pre- or post-Duchamp".

In the case of MVRDV's New Orleans Bend House, the playful break with convention was not accomplished without considerable debate. "When you have a federal government that doesn't invest in its levees, that makes people's land completely worthless, that makes its own citizens insanely poor, you need a design that makes a protest, that rises up and says, What is going on here?" Maas said. "But in discussions with Brad and the others, we kept asking: Yes, but can we show that explicitly? Can we come out with that? It's going to look ironic! How can you be ironic in the face of disaster? Will the American people be angry?"

"But even in the most tragic circumstances," Maas went on, "there is often a moment of irony. Well, is it irony? Or is it really more like …?" He paused, at an uncharacteristic loss for words. "There is this beautiful German word, Trost. It means empathy, or solace, or maybe consolation. I think that is what our building meant to express. You know, if the waters are going to come, let them come. Let's do it. Let's just turn and face it."

Resources for Reference

http://www.mvrdv.nl/_v2/

http://www.mvrdv.nl/_v2/projects/010_villavpro/index.html

http://www.nytimes.com/2005/04/10/arts/design/10ouro.html

http://www.oma.eu/

http://www.nytimes.com/2008/06/08/magazine/08mvrdv-t.html?pagewanted=1&fta=y

Section 3
Tips for Translation

Translating appositive clauses 同位语从句的翻译

同位语从句的一般译法，是将从句与先行词分开，翻译成独立的句子从而使译文结构清楚，更符合汉语语言习惯。

1. People came to understand a structural behavior *that stiffness and strength of sheet-like elements can be obtained not only by increasing their thickness...*

人们开始了解一种新的结构现象，那就是不需要增加厚度，就可以使纸一样薄的构件获得刚度和强度……

2. We owe to the greatest of all mathematicians, Karl F. Gauss (1777-1855), the discovery *that all the infinitely varied curved surfaces we can ever find in nature or imagine belong to only three categories, which are dome- like, cylinder-like, or saddle-like.*

我们必须把下面的发现归功于最伟大的数学家高斯：迄今我们能够在自然界中发现或想象出来的无数不同的曲面都属于三个类别，也就是穹顶状、柱状和马鞍状。

3. He takes the position *that not enough is yet known about how to design non-treelike cities to provide definite answers.*

他的立场是：目前关于如何设计非树形结构的城市了解得还不够，所以还无法给出明确的答案。

Unit 14

Section 1
Intensive Reading

Modern Housing Prototypes
Roger Sherwood

Part V

Double-Orientation Unit, Open-Ended

While single-orientation units are suitable for buildings with double-loaded corridors that open to each side and for hillside housing or single-loaded corridor buildings that turn their backs upon some undesirable site condition such as a highway or a northern exposure, housing units with a double orientation are far more common. Probably stemming from the common sense advantage of repeating units while maintaining maximum exterior surface, this system of placing open-ended units side by side is perhaps the oldest form of collective urban housing. A dwelling unit that is open at each end has many organizational options. If the unit is very deep, light is minimal and the open ends are not much of an advantage.

Figure 14-1

O. M. Ungers' Green Belt South housing of 1965 (Figure 14-3) or the Backen, Arrigoni, and Ross project in Tustin (Figure 14-4) are good examples of very deep units. In each, unit is so long that some auxiliary means of lighting the interior has been used. With Ungers, a parallel open slot lets light into the four-story building, while in the Tustin project a system of interior courtyards is used, resulting in a one-story building.

By comparison, units such as Lurcat's rowhouses at the Vienna Werkbund Exposition (Figure 14-5) do not have a light problem because they are so shallow. But because the rooms are small, core elements come to an outside wall and the stairs are actually attached to the exterior as a separate element. So there are general criteria for optimum depth: shallower units could very well become single-orientation types, deeper units have to find some other means of introducing light, such as interior courtyards (which are unsuitable for high-rise buildings). Optimum widths and depths is also a function of building requirements: room sizes, stairs, and so on.

The open-ended slot requires open space outside the unit at each end and usually some means of providing privacy——a garden wall, for example——except where the unit is well off the ground. Access to this type can be from either end or, in the case of multistory buildings, from within, making internal skip-stop corridor systems mandatory. Walk-up units, which were especially popular in Europe before the postwar proliferation of high-rise building——Siemensstadt, for example (Figure 14-6)——also give access at an interior point.

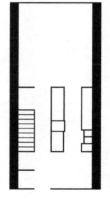

Figure 14-2

Generally, the double-orientation type is at least a two-story unit, so the architect must consider where the stair, kitchen, and bath can be put. Basically, the types may be classified as either transverse (stair perpendicular to the long axis of unit) or longitudinal (stair parallel to the long axis). Following are a few examples of the double-orientation types.

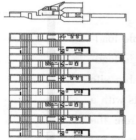

Figure 14-3 Green belt, South housing, Zollstock, Germany. O. M. Ungers, 1965.

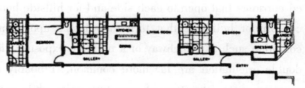

Figure 14-4 Patio housing, Tustin, California. Backen, Arrigoni, and Ross, 1069.

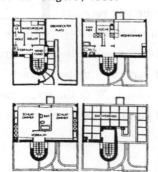

Figure 14-5 Rowhouses, Werkbund Exhibition, Vienna. Andre Lurcat, 1932.

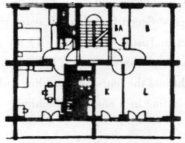

Figure 14-6 Siemensstadt housing, Berlin. Fred Forbat, 1930.

Double-orientation unit, open-ended; interior stair and core, longitudinal stair. Sometimes called a "dumbbell" plan because of its form——a void at each end and a concentration of parts in the middle——this type positions the major living spaces to the outside, where an opening to private outdoor space is a possibility, and keeps the core elements, including the stairs, on the interior. The dumbbell plan rowhouse has a tradition dating back to medieval times and in most Western cities was probably the most common form of housing until the invention of the rigid structural frame. Historic examples are wide-ranging: Scamozzi's Procuratie Nuove of the seventeenth century in Venice (Figure 14–7) is entered from an interior courtyard via stairs, with major living spaces facing the Piazza San Marco and sleeping spaces opening to the garden and the Grand Canal——palatial quarters ingeniously planned, of incredible beauty. On the other hand one may find a rowhouse from Baltimore of the nineteenth century, which is typical of urban housing in the eastern part of the United States prior to 1920 or so (Figure 14–8). The dumbbell plan is popular in the United States because building codes allow interior kitchens and baths. Other examples include the rowhouses at Reston (Figure 14–9) and those at Roehampton by the London County Council (Figure 14–10).

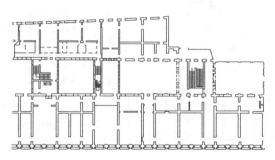

Figure 14–7 Procuratie Nuove, Venice. Vincenzo Scamozzi, seventeenth century.

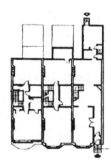

Figure 14–8 Typical townhouse, Baltimore. Nineteenth century.

Figure 14–9 Rowhoues, Reston, Virginia. Whittlesey and Conklin, 1964.

Figure 14–10 Rowhouses, Roehampton. London County Council, 1952.

The typical early twentieth-century walk-up housing consisted of a dumbbell plan that was entered from an interior hallway. Even though European building codes tend to require that kitchens have exterior windows, a dumbbell type of plan usually results. Examples are Sert's Peabody Terrace at Harvard (Figure 14–11) and Siemensstadt (Figure 14–6), the huge project outside Berlin of the 1930s. There one finds many different buildings done by many different architects, but all are just minor variations of the same unit type——a situation probably encouraged, in Germany at least, by Mies van der Rohe's block at the Stuttgart Weissenhof exhibition of 1927 (Figure 14-12). This is the type of walk-up unit that was used in Germany to the practical exclusion of all else for almost two decades. The walk-up unit with a dumbbell plan is also popular in England, the King Street project by Morton, Lupton, and Smith (Figure 14-13) perhaps being representative of recent rowhousing there.

Figure 14-11 Peabody Terrace, Cambridge, Massachusetts. Sert, Jackson, and Gourley, 1964.

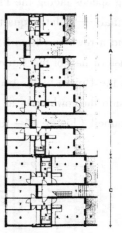

Figure 14-13 King Street housing, London. Morton, Lupton, and Smith, 1967.

Figure 14-12 Weissenhof, exhibition housing, Stuttgart, Mies van der Rohe, 1927.

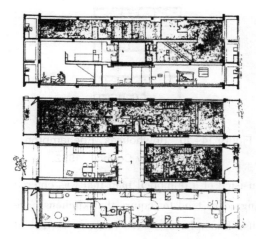

Figure 14-14 Unite d'Habitation, Marseilles. Le Corbusier, 1952.

The dumbbell plan is not restricted to use in rowhouse or walk-up situations. It also has wide application in high-rise buildings, particularly slabs. Le Corbusier's Unité d'Habitation (Figure 14-14), a building that has been repeated in slightly differing versions in France and Germany and has been widely copied almost everywhere, is the perfect example of a multistory dumbbell plan. A double-loaded, skip-stop corridor gives access to a two-level unit with kitchen, dining, and living area at entry level and bedrooms and bath above. Here the core elements, including the stair, are interior, although the stair rises from a double-height living room.

Words and phrases

1. open-ended   *adj.*   末端开口的
2. auxiliary mean   辅助设施
3. slot   *n.*   缝，槽
4. optimum   *adj.*   最适宜的
5. mandatory   *adj.*   强制性的
6. proliferation   *n.*   增殖，大量出现
7. longitudinal   *adj.*   经度的，纵向的
8. dumbbell   *n.*   哑铃
9. palatial   *adj.*   富丽堂皇的
10. hallway   *n.*   门厅，回廊，走廊

Section 2
Extensive Reading

What Is Green Building?

It's a concept with quite the buzz. Everyone is talking about it, from Wired magazine, to Elle, to Vanity Fair. It's a topic so popular, that a Google search yields over 8 million hits.

It's how YOU can make a difference in the future of your family, your city, your planet. And it's a great way to think globally and act locally.

It Started with Sustainable Development

In 1987 the United Nation's World Commission on Environment and Development, known as the Brundtland Commission, met to create a vision called "Our Common Future" that was based upon sustainability. From this effort, the definition of sustainable development emerged as:

"meeting the needs of the present without compromising the ability of future generations to meet their own needs."

Sustainable development measures success in terms of economic, environmental, and social benefits. The building industry expanded on this concept, and applied it to "the built environment," creating the term sustainable building. The term sustainable building is used interchangeably with green building. Its purpose is to reduce the adverse human impacts on the natural environment, while improving our quality of life and economic well-being.

The Impact of the Building Industry

Buildings and development affect water quality, air quality, and ecosystems, impacting human healthy and our quality of life. In addition to environmental impacts, buildings have a large economic footprint. Buildings represent more than 50 percent of the nation's wealth, and the U.S. construction market comprises 13% of U.S. GDP, and building-related fields employ 10 million people (2003 U.S. DOE Buildings Energy Databook). The U.S. construction market is also responsible for:

- √ 39% of total energy use
- √ 39% of municipal solid waste
- √ 35% of greenhouse gas emissions
- √ 40% of all raw materials, including
- √ 25% of timber harvests
- √ 12% of potable water withdrawal

To remain competitive and continue to expand and produce profits in the future, building industry professionals are learning to address the environmental, social and economic consequences of their industry. Through careful planning, we can substantially reduce the adverse impacts of the built environment. Some strategies can actually improve degraded environments and increase the comfort and productivity of building occupants. Sustainable building is an integrated approach that promotes environmental quality, economic vitality, and social benefit through the design, construction and operation of the built environment.

Green Building Today

Green building applies principles of resource and energy efficiency, healthy buildings and materials, and ecologically and socially sensitive land-use to achieve "an aesthetic sensitivity that inspires, affirms, and ennobles".

Green building requires an integrated, multi-disciplinary design process and a "whole-building" systems approach that considers the building's entire life-cycle (from planning, design, and construction to operation and maintenance, renovation, and demolition or building reuse). Together, these provide the means to create solutions that optimize building cost and performance.

Resources for Reference

http://en.wikipedia.org
http://www.seattle.gov/DPD/
http://www.eu-greenbuilding.org/

http://www.seattle.gov/dpd/GreenBuilding/OurProgram/Overview/WhatisGreenBuilding/default.asp

Section 3
Tips for Translation

Translating participial phrases used as attributive 用作定语的分词短语的翻译

英语中有现在分词短语（present participial phrases）和过去分词短语（past participial phrases）两种形式。这两种分词短语在句子中均可作为定语使用，可以前置，亦可后置。在翻译中一般也翻成定语。

1. Too many designers today seem to be yearning for the physical and plastic characteristics of the past, instead of searching for the abstract *ordering* principle which the towns of the past happened to have, and which our modern conceptions of the city have not yet found
看起来如今太多的设计师都在怀念过去时代的物质形态特征，而不是寻求深层的控制性原则，这些原则是过去的城镇正好拥有的，也正好是我们在构想现代城市的时候尚未发现的。

2. The load capacity obtained by such a flimsy piece of material through its folds is amazing: a sheet of paper *weighing less than one-tenth of an ounce* may carry a load of books two or three hundred times it own weight!
通过折叠而使如此之薄的材料所获得的承载力令人惊讶：一张不过十分之一盎司重的纸却可以支撑其自身重量两百到三百倍的书。

3. This article shows how the apparently conflicting intentions of the architect and the engineer can be fused into a *unified* creative process.
文章表明：建筑师同结构工程师之间看起来格格不入的想法是能够被融入一个统一的创作过程的。

4. Frampton's essay was included in a book Essays on Postmodern Culture, *edited* by Hal Foster, though Frampton is critical of postmodernism.
弗兰普顿的文章收录在哈尔·福斯特编著的《后现代文化评论》一书中，尽管弗兰普顿对后现代主义是持批判态度的。

5. There are a lot of students *staying up all night* for the final project in the classroom.
教室里有不少学生在为毕业设计熬夜。

6. The college is a group of departments, *each serving a special purpose*: the department of architecture, the department of urban planning, the department of landscape architecture, and the like.

学院包括几个系，每个系的专业不同，如建筑系、城市规划系、景观建筑系等。

　　在例句 5 和 6 的翻译中，为了译文的顺畅，把分词短语可转换成分句或补语翻译。

Unit 15

Section 1
Intensive Reading

Modern Housing Prototypes
Roger Sherwood

Part VI

Double-orientation unit, open-ended; exterior kitchen, longitudinal stair.

Perhaps a more common version of the longitudinal stair arrangement, and one popular in Europe, brings the kitchen to the outside; either the bath for the bedrooms is above the kitchen or another core or service wall is introduced on the interior. Examples of this include the Milton Road project (Figure 15 – 4) and Siedlung Halen by Atelier 5 (Figure 15 – 5), both self-contained rowhouses or terrace houses, or a walk-up situation also from the Milton Road project by the Borough of Haringey (Figure 15 – 4). This unit type is also used in high-rise slabs but again, because it is a two-story unit, it is limited to skip-stop corridor arrangements. Single-loaded and double-loaded corridor arrangements are feasible, with the kitchen usually at the entrance level; in the case of the single-loaded type with access gallery, the kitchen gets light from the gallery. Examples of the double-loaded type include Schmiedel's apartments in Germany (Figure 15 – 6) and the van den Broek and Bakema Hansa block in Berlin (Figure 15 – 7).

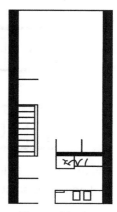

Figure 15 – 1

Single-loaded versions include the L'Aero Habitat development by Bourlier and Ferrier in Algiers of 1950 (Figure 15 – 8), Werner Seligmann's hillside housing in Ithaca, New York, of 1972

(Figure 15-9), and Swiss Cottage in London by Douglas Stephen with Koulermos and Forrest (Figure 15-10).

Double-orientation unit, open-ended; exterior kitchen, transverse stair.
This is the most common open-ended unit. Although usually wider than a unit with a longitudinal stair, several advantages result. First, a clear circulation zone along one wall is defined by the stairs and other core functions along the opposite wall. Circulation in the living room is now along the side of the space, and from the

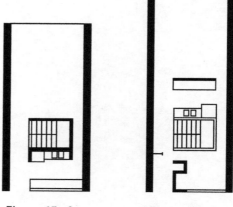

Figure 15-2   Figure 15-3

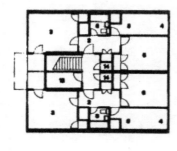

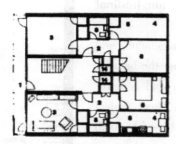

Figure 15-4 Milton Road rowhouses, London. District of Haringey, 1967.

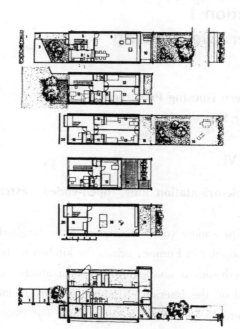

Figure 15-5 Siedlung Halen, Bern. Atelier 5, 1959.

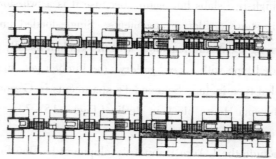

Figure 15-6 Apartments, Germany. Schmiedel and Zumpe, 1960

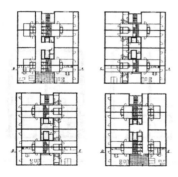

Figure 15-7 Hansaviertel tower, Berlin. Van den Broek and Bakema, 1956.

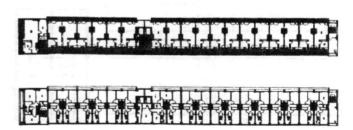

Figure 15-8 L'Aero Habitat, Algiers. Bourlier and Ferrier, 1950.

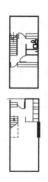

Figure 15-9 Hillside housing, Ithaca, New York. Werner Seligmann, 1972.

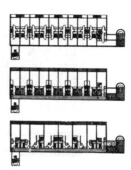

Figure 15-10 Swiss Cottage, London. Douglas Stephen, 1960.

entrance one can see down the hallway into the living area, which gives the impression of one continuous space throughout the floor. If the stair is pulled back slightly from each side wall, allowing enough space to move past the stair, the kitchen can serve the living and dining area past the stair without using the main hallway. This unit satisfies the European preference for exterior-fronting kitchens and forms a larger space in the living and dining area where it is most useful. Upstairs, unless the plumbing stack can be manipulated so that the bathroom is on the interior, valuable exterior surface is taken up with a space requiring only minimal light and air. Either single-run or return stairs can be used, and different minimum unit widths, of course, result. Low-rise examples include the Flamatt terrace houses in Bern by Atelier 5 (Figure 15-11) and their famous Siedlung Halen, also in Bern (Figure 15-5).

This is a common type for use in high-rise buildings. The serpentine slab of Affonso Reidy in the Pedregulho development in Rio de Janeiro (Figure 15-12), the Billardon slab at Dijon by Beck (Figure 15-13), and Womersley's Park Hill project (Figure 10-2) are three examples.

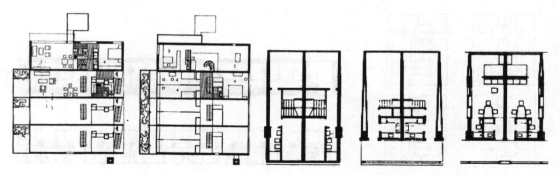

Figure 15-11 Flamatt terrace housing, Bern. Atelier 5, 1960.

Figure 15-12 Pedregulho housing, Rio de Janeiro. Affonso Reidy, 1950.

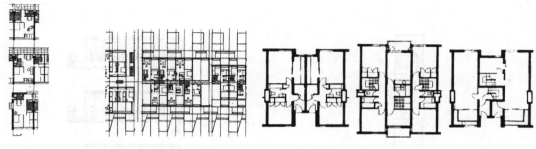

Figure 15-13 unité Billardon, Dijon. Pierre Beck, 1954.

Figure 15-14 Terrace apartments, Germany. Schroder and Frey, 1959

Figure 15-15 Edith Avenue housing, Durham, England. Napper, Errington, collerton, Barnett, and Allot, 1961

Double-orientation unit, open-ended; interior kitchen, transverse stair.

This version of the dumbbell type, with stair and other core elements on the interior, comes in many variations, some with stair, kitchen, and bath on the same side, some with the kitchen opposite the stair. Examples of the latter arrangement include the terraced walk-up flats of Schroder and Frey (Figure 15-14). Sometimes the transverse stair and the kitchen are together, a type common in row-house applications——for example, the Edith Avenue housing project of 1961 in Durham (Figure 15-15), and the Amis and Howell houses in Hampstead of 1956 (Figure 15-16).

Few high-rise buildings seem to use this type. However, the Bresciani project, Quinta Normal in Santiago, Chile (Figure 15-17), uses an interlocking system with living and dining areas and kitchen taking up two bays to one side of the corridor at the lower floor and the bedrooms in an open-ended arrangement above but in just one bay. This system would be applicable for high-rise slabs as well.

The double-orientation, dumbbell unit plan is impractical for very shallow buildings where there is seldom room for the interior core. In Lurcat's Vienna project, for example, the core has to come to

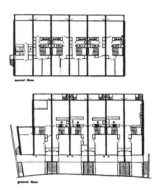

Figure 15-16 Rowhouses, Hampstead. Amis and Howell, 1956.

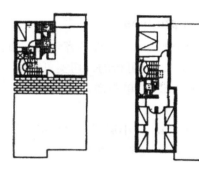

Figure 15-17 Quinta Normal, Santiago. Carlos Bresciani, 1960.

the outside, although here each unit is three floors high. With Neave Brown's Fleet Road project (Figure 15-18), a similar situation occurs: a three-bedroom maisonette has kitchen and bath fronting the gallery, but, because of the limited area on any on floor, a peculiar mix of spaces results in which the dining area is separated form the living room, bedrooms are on both floors, and toilet facilities are of necessity duplicated. For high-rise building, this type is probably not suitable: a very narrow building would be structurally unstable if higher than a few floors unless the buildings were warped for added lateral support. The Smithsons' curved slab project of the 1950s (page 132) was presumably developed just for this reason.

Another double-orientation, open-ended unit type that is widely used expands laterally; bedrooms, rather than being upstairs, take over the adjacent bay, so that the entire apartment is on one floor but in two bays. In a two-story unit, there is an overlapping of bays so that bedrooms above would be over the living rooms of both units below. The single-floor version is typical of most walkup housing or noncorridor types of high-rise buildings. The Decoppet, Veuve, Aubry, and Mieville project in Lausanne, Switzerland (Figure 15-19), is a good example of this type.

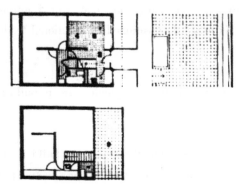

Figure 15-18 Fleet Road terrace housing, London. Neave Brown, 1968.

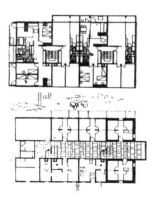

Figure 15-19 Apartment block, Lausanne. Decoppet, Veuve, Aubry, and Miéville, 1959.

Words and phrases

1. self-contained *adj.*   设备齐全的，独立的
2. feasible *adj.*   可行的，切实可行的
3. single-run *n.*   单跑楼梯
4. bay *n.*   开间
5. maisonette *n.*   小房屋，出租房间
6. overlap *v.*   交搭，叠盖

Section 2
Extensive Reading

Why Build Green?

Building green isn't just about improving your health or saving the planet. Today's techniques can also save you money and make your projects more marketable.

Green building techniques reduce energy and water use, improve indoor air quality, are sensitive to site development issues, incorporate environmentally friendly building materials, and more. Building green benefits:

√ design and construction professionals
√ building owners and occupants
√ Seattle's communities and the environment at large

Benefits to Design and Construction Professionals

Seattle's Comprehensive Plan predicts annual construction growth through 2009 will include over 13 million square feet of residential (single and multifamily) construction, and over 5 million square feet of commercial and industrial construction. With Seattle positioned as a national leader in green building, opportunities to go green abound.

Developers and professionals in design and construction benefit from employing green building practices in many ways:

√ **Market advantage.** As demand increases for green buildings and the firms that build them, those that demonstrate expertise in green building will be natural candidates for prospective projects.

√ **Goodwill.** Becoming known as a "green firm" can enhance your business's reputation in the community and help you recruit and retain a superior workforce.

√ **Proactive regulatory stance.** Employing design and construction practices that routinely exceed code means your firm is primed for future increases in code stringency, reducing ramp-up and retooling costs.

Benefits to Building Owners and Occupants

Contrary to popular belief, green building may not always cost more. If it does, a price premium of 1% ~ 3% can also pay back over time, and provide the following benefits:

√ **Increased property value.** As a superior product, a green building can yield increased value to the owner. For example, *Green Value: Green Buildings, Growing Assets*, a study of 11 green buildings in the U.S. and Canada, found that——in addition to the payback efficiency measures provide over time——a "green" designation can also increase a building's market value as assessed by appraisers and investors.

√ **Increased return on investment.** Up-front investments in green designs and technology can yield financial paybacks over the life of the building in the form of lower utility bills and reduced operating and maintenance costs. For example, a 2003 report, The Costs and Financial Benefits of Green Buildings, prepared for California's Sustainable Building Taskforce, analyzed both first costs and 20 year operational costs and savings of 30 completed LEED projects. On average the additional first cost was 2%, or $3.00 ~ $5.00/square foot. However, the total 20 year net present value of improvements attributed to energy, emissions, water, waste and commissioning is $11.98.

√ **Enhanced occupant health and well-being.** Access to daylight and fresh air and creating an environment free of toxins and irritants helps protect building occupant health and enhances well-being. This too can result in a financial benefit: the same study Referenced above reports a benefit to health and productivity that dwarfs even the reduced resource use savings. Adding the value of improved tenant health and productivity, the 20 year net present value for LEED Certified and Silver buildings is $48.87 and for Gold and Platinum $67.31.

√ **Boost occupant productivity.** Numerous studies show the productivity benefits of fresh air and natural light in buildings, and conversely, the negative impact of poor indoor environmental quality. The Heschong-Mahone Group found that the quality of light and air in a workspace can affect worker productivity by up to 20% either positively or negatively.

√ **Increased sales and leasing potential:** Green building features can help close the deal. For example, the Brewery Blocks in Portland, OR leased out more quickly and at an enhanced rate per square foot than comparable space in the Portland market, even as a depressed economy was resulting in a net loss of leased space across the city. (See the Brewery Blocks case study for more.)

Benefits to Seattle's Communities and the Environment

Many benefits of green building don't register on the bottom line of the developer, design or construction professional or building owner or occupant. But the benefits are still tangible and valuable:

√ Enhancement of community and local economy
√ Public health benefits
√ Protection of threatened and endangered species
√ Supporting sustainable resource use

Resources for Reference

http://www.seattle.gov/dpd/
http://www.greenbuilding.com/
http://www.seattle.gov/dpd/GreenBuilding/OurProgram/WhyBuildGreen/default.asp

Section 3
Tips for Translation

Translating participial phrases used as adverbial 用作状语的分词短语的翻译

现在分词短语（present participial phrases）和过去分词短语（past participial phrases）在句子中也经常作为状语使用。

1. Dr. and Mrs. Soleri made a life-long commitment to research and experimentation in urban planning, *establishing the Cosanti Foundation, a not-for-profit educational foundation*.
索拉里夫妇毕生致力于城市规划的研究和实验，并设立了非营利的 Cosanti 基金会。

2. To combat the glass box future, many valiant protests and designs have been put forward, *all hoping to recreate in modern form the various characteristics of the natural city which seem to give it life*.
为了抗击玻璃盒子的未来，人们提出了不少大胆的主张和设计，希望在现代的形式中再创那些赋予自然城市以生命的各种特质。

3. Almost all scientific works on Sitte until today focused exclusively on his theory of urban planning, *neglecting a systematic analysis of his voluminous estate*, which consists of 60 writings on architecture and urban planning, 60 writings on music, painting, art history and arts and crafts, 19 writings on pedagogy, numerous letters, 25 architectural and 17 urban design projects.
迄今为止关于希特的学术论文都是关于他的城市规划理论的，但却忽略了对他丰富学术成果的系统分析，包括建筑学和城市规划方面的 60 部著作，关于音乐、绘画、艺术史、艺术和手工艺的 60 部著作，19 部教育学著作，大量的信札，以及 25 个建筑和 17 个城市规划项目。

4. *Though not so commonly used as rectangular grids*, skew grids have, beside aesthetic qualities, the structural and economic advantage of using equal length beams even when the dimensions of the grid are substantially different, *thus distributing more evenly the carrying action between all the beams.*

尽管不像正交网格那么使用得多，斜交网格除了在美学上的特征之外，当网格在两个方向上尺度差异相当大的时候，相比较于采用同等长度的梁，斜交网格能更均匀地在所有的梁之间分布荷载，从而具有结构和经济上的优势。

分词短语作状语的时候，往往在句子中表示时间、原因、结果、条件、方式、伴随等各种情况。在翻译中可以添加相应的连接词，把句子的不同的含义表达出来，也使汉语译文更加顺畅。

Unit 16

Section 1
Intensive Reading

Modern Housing Prototypes
Roger Sherwood

Part VII

Building Types

The ways in which the various dwelling units can be combined into different building forms are a function of the special characteristics of the building——site, orientation, height, and so on—— and the circulation system used. Because the ways in which units may be collected together are limited by building regulations, construction practices, and cultural preferences, different housing types occur in some countries while not in others. For example, United States fire codes, until very recently, required an exit from each floor of an apartment and so eliminated skip-stop sections like the typical Unité of Le Corbusier. In some countries, such as France and Brazil, multiple fire stairs are not required; and in Chile five-story walk-ups are allowed. Sometimes a particular housing form may result from a tradition of similar housing; the widespread construction of four-story walk-up buildings in Germany, the gallery-access maisonette in England, or hillside housing in Switzerland. Although absolute comparison of housing from country to country would have to take into consideration the differences in building regulations, construction practices, and national traditions, comparison is possible on the basis of unit and building types. It is not necessary to understand all about building in a particular place to be able to analyze a particular building, to classify it organizationally, and to identify its unique features and concepts. Without a comprehensive understanding of building practices in every country——an unlikely knowledge——comparison on any other basis seems all but impossible.

Building forms resulting from the collecting together of many units into a single building are closely tied to a few possible circulation options. If a community of dwellings is seen as simply many individual houses, each hooking on to an access system, then only a few systems emerge.

Private Access (Figure 16-1)

Here there is private entrance and private internal vertical circulation. Height is limited by most building codes to two or three stories. Units cannot be stacked vertically and the idea is restricted to rowhouses, detached houses, or terrace houses. Neave Brown's five houses on Winscombe Street in London (Figure 16-3) are examples of this type.

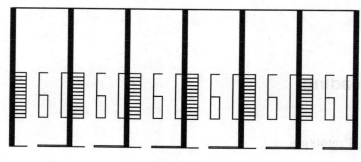

Figure 16-1

Multiple Vertical Access (Figure 16-2)

This type can be built up to five stories without elevators in some countries, but more often three stories is the limit for walk-up multiple-access buildings. Taller buildings can be developed with the use of elevators, but the expense of repeating elevators is an obvious limitation. Multiple vertical access buildings were very common in Europe before World War II and the subsequent rapid construction of high-rise buildings. Usually, each access stair serves two to four units per floor with semiprivate entrance to each apartment. Since the system permits vertical stacking, it becomes a kind of vertical rowhouse, or rowhouses stacked upon rowhouses. In the United States, where multiple fire exits are required in housing over two floors, this type has never developed. Typical European examples include the Wellhausen project in Hamburg of 1967 (Figure 16-4), where the access core is treated as a separate, external element consisting of a stair for the three-story block and a stair and elevator for the six-story block, and the Candilis, Josic, and Woods walk-ups at Nimes of 1961 (Figure 13-13), where the stair for a five-story walk-up is the connecting element between apartment blocks, generating a kind of continuous, repetitive build-

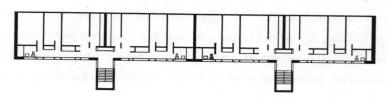

Figure 16-2

ing. In the typical housing in Germany of the 1920s and 1930s——Siemensstadt, for example (Figure 14–6)——the access stairs are internal, between units, with only minor articulation indicating the position of the stair on the exterior.

If the vertical access core is greatly extended and centralized, the result is a tower, which may be described as a group of units hooked together along a vertical street. There are countless variations to the tower plan, but it usually consists of several units per floor. Because normally light is required from all sides, a freestanding building (point block) usually results, such as Mies' Lake Shore Drive apartments in Chicago, a twenty-nine story building (Figure 12–3). Sometimes, however, the tower connects to other, lower buildings like the four Ancker and Gate towers at Vallingby (Figure 16–5). At other times the tower is simply multiplied and connected together to form the continuous building type like the Bandel blocks of 1967 (Figure 16–6). Although different types of units may be used with multiple vertical access buildings, the walk-up situation is probably better suited for the double-orientation type, and in this respect it is like a rowhouse. With the tower, the single-orientation unit type is more typical, with a double-orientation, 90° unit at the corner, although again there are countless possible variations.

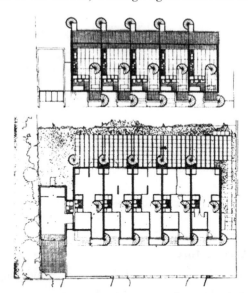

Figure 16–3   Winscombe Street houses, London. Neave Brown, 1968.

Figure 16–4   Apartment block, Hamburg. George and Michael Wellhausen, 1967.

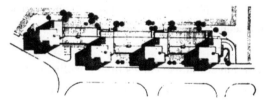

Figure 16–5   Towers, Vallingby, Sweden. Ancker and Gate, 1953.

Figure 16–6   Apartment blocks, Britz-Buckow-Rudow, Germany. Hannskarl Bandel, 1967.

Corridor Buildings

The term "slab" implying a tall, long building, is commonly used to describe corridor buildings, although a corridor system is not limited to high-rise buildings. Dwelling units in a slab simply align along a continuous corridor that has periodic connections to the ground. Building height and vertical access requirements are a function of building regulations and varying economic considerations such as elevator cost and other mechanical services. However, slab heights vary widely and any optimum condition is more the result of local building conditions.

Corridor buildings come in two basic types, single-loaded and double-loaded, and there are many variations of each. Some have corridors every floor, others have corridors every second, third, or even fourth floor. Some have corridors occurring at different positions in the section at different levels.

Words and phrases

1. hillside housing 山地住宅
2. hook on to 追随
3. detached house 独立式住宅
4. terrace houses 退台式住宅
5. multiple-access 多个出入口的
6. semiprivate *adj.* 半私密的
7. align *v.* 排成直线
8. periodic *adj.* 周期的

Section 2
Extensive Reading

Words and Buildings: A Vocabulary of Modern Architecture

Since the eighteenth century, architects had talked about "volumes" and "voids"——and they occasionally used "space" as synonym: Soane, for example, referred to "void spaces", and to the need in devising the plan to avoid "loss of space". Although 'space' is often still used in this sense, it was to convey something more that "space" was adopted by modernist architects, and it is with these superposed meanings that we shall entirely concern ourselves here.

Much of the ambiguity of the term "space" in modern architectural use comes from a willingness to confuse it with a general philosophical category of "space". To put this issue slightly differently, as well as being a physical property of dimension or extent, "space" is also a property of the mind, part of the apparatus through which we perceive the world. It is thus simultaneously a thing within the world, that architects can manipulate, and a mental construct through which the mind knows the world, and thus entirely outside the realm of architectural practice (although it may affect the way in which the results are perceived). A willingness to connive in a confusion between these two unrelated properties seems to be an essential qualification for talking about architectural space. This confusion is present in most of what is said about architectural space; it finds its expression in the commonly held belief that architects "produce" space——a belief implicit in the statements of Denari and Lasdun quoted at the opening of this entry. It was part of the purpose of Henri Lefebvre's The Production of Space (1974) to expose the problem created by this distinction between space conceived by the mind and the "lived" space encountered by the body; Lefebvre's book, the most comprehensive and radical critique of "space", calls into question almost everything about space within architecture described in what follows——but despite its force, it has had, as yet, little impact upon the way space is still customarily talked about within architecture.

Resources for Reference

http://en.wikipedia.org

http://www.amazon.com/

Adrian Forty. *Words and Buildings*: *A Vocabulary of Modern Architecture*. 2004: Thames & Hudson

Section 3
Tips for Oral Presentation

Transitions

Transitions are an integral part of a smooth flowing presentation, yet many speakers forget to plan their transitions. The primary purpose of a transition is to lead your listener from one idea to another. The following are some examples of transitions that work well:

- Bridge words or phrases (furthermore, meanwhile, however, in addition, consequently, finally).
- Trigger transition (same word or idea used twice: "a similar example is …").
- Ask a Question ("How many of you …. ?")
- Flashback ("Do you remember when I said …?")
- Point-By-Point ("There are three points …The first one is…The second one is…etc.")
- Add a Visual Aid as a Transition——Many times it may be appropriate to add a visual between your regular visual aids for the sole purpose of a "visual" transition. Many times a clever car-

toon used here can add some humor to your presentations.
- Pausing (Even a simple pause, when effectively used, can act as a transition). This allows the audience to "think" about what was just said and give it more time to register.
- Use Physical Movement (The speaker should move or change the location of their body. This is best done when you are changing to a new idea or thought).
- Use a Personal Story The use of a story, especially a personal one is a very effective technique used by many professional speakers. (Used effectively, it can help reinforce any points you made during your presentation).
- Use the PEP formula (Point, Example, Point) (This is a very common format used and can also be combined with the use of a personal story. Make sure stories or examples you use help reinforce your message).

Using transitions

Three common mistakes made when using transitions:

- The most common mistake people make is that they DON'T use transitions at all. Transitions help your information flow from one idea to the next.
- The second most common mistake is using transitions that are too short. Not enough time is spent bridging to the next idea. This is extremely important when changing to a new section of ideas within your presentation.
- The third most common mistake is that people use the same transition throughout the presentation. This becomes very boring after a short while. Try to be creative with your transitions.

Transitions and the Team Presentation

Transitions become extremely important when a team presentation is involved. The transition from one speaker to the next must be planned and skillfully executed. Each speaker should use a brief introduction of the next topic and speaker as part of this transition.

Resource for Reference

(http://www.ljlseminars.com/transit.htm)

Unit 17

Section 1
Intensive Reading

Modern Housing Prototypes
Roger Sherwood

Part VIII

Single-Loaded Corridor Systems
Buildings of this type generally open to the side away from the corridor and hence are commonly used where there may be a preferred view or orientation or some undesirable site condition that the unit can, in effect, turn its back to. A corridor-every-floor system usually results in a building made up of single-orientation units; an alternating corridor system often results in two-level or marsonette unit types, with both single and double orientation. Where the climate permits, the corridor can remain open (gallery access) and becomes a kind of street in the air, a concept evolved in 1919 by Brinkman in the Spangen Quarter in Rotterdam (Figure 17-4) and employed in postwar English housing such as Park Hill (Figure 10-2). The Narkomfin collective housing project in Moscow by Ginzburg of 1928 (Figure 17-5) is an enclosed version of an alternate-level gallery-access system.

Single-loaded system; corridor every floor (Figure 17-1).
Examples of this type include the Bergpolder slab in Rotterdam of 1933 by the team of van Tijen, Maaskant, J. A. Brinkman, and van der Vlugt (Figure 17-6), a very early experiment in high-rise housing; the Billardon slab at Dijon by Beck of 1954 (Figure 15-13); and Alvar Aalto's apartments at Bremen of 1958 (Figure 12-6).

Single-loaded system; corridor every second floor (Figure 17-2).
This popular type was frequently used in postwar, low-rise housing. It consists of maisonettes off an

access gallery with bedrooms above, often over the corridor. Stirling and Gowan's Preston housing at Lancashire of 1961 (Figure 12 – 8) demonstrates the type: three-story buildings with private entrance to a lower level and an access gallery for the upper maisonettes. Stirling's Runcorn housing (Figure 17 – 7) is perhaps an evolutionary development of the same scheme, with the building now five stories high and a gallery at the third floor. Here the maisonette on the bottom two floors has private entrance at ground level, the gallery gives access to the maisonette on the next two floors, and stairs give access to the flat on top, which extends over the gallery. Brinkman's Spangen Quarter (Figure 17 – 4) is a very early example of this type. Here the gallery, really an independent structure, services upper maisonettes while independent stairs and private entrance give access to the two lower units in a four-story building. Le Corbusier's Immeuble Villas projects (Figure 12 – 8) are more extravagant: two corridors side by side, one service and one public, give access to a huge two-story unit with a double-height living room and large terrace. The same idea is also used in much taller buildings. For instance, the L' Aero Habitat slab of Bourlier and Ferrier in Algiers of 1950 (Figure 15 – 5)——a thirteen-story building and a series of slabs, one placed perpendicular to a steep slop——and Villaneuva's El Paraiso slabs in Caracas of 1956 (Figure 17 – 8).

Single-loaded system; corridor every third floor (**Figure** 17 – 3).
The more unusual types of single-loaded, alternate-level corridor buildings position a corridor every third floor with stairs up or down to the units that are not at the corridor level.

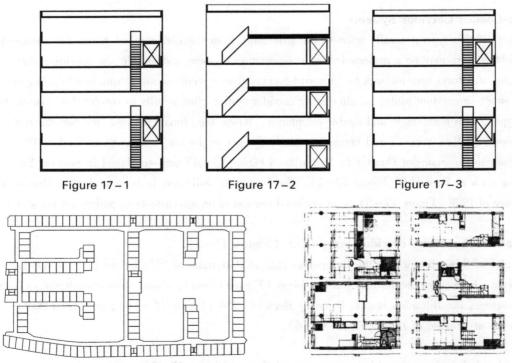

Figure 17 – 1   Figure 17 – 2   Figure 17 – 3

Figure 17 – 4   Spangen Quarter, Rotterdam. Michiel Brinkman, 1919.

Figure 17 – 5   Narkomfin Apartments, Moscow. Moses Ginzburg and I. Milinis, 1928.

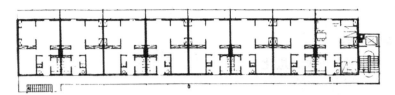

Figure 17-6 Bergpolder apartments, Rotterdam. Van Tijen, Maaskant, J. A. Brinkman, and van der Vlugt, 1933.

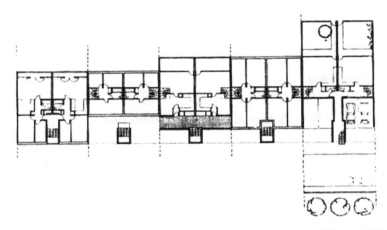

Figure 17-7 Rowhouses, Runcorn, England. James Stirling, 1968.

Sometimes there are maisonettes at the corridor level with a smaller apartment below, sometimes there are larger units below. This is strictly a low-income housing type except where the maisonette is used, and it is typical of high-density, low-income public housing such as Park Hill (Figure 10-2). Slabs with a single-loaded corridor only every fourth floor are quite unuaual because few building codes allow such a considerable inconvenience. However, this kind of building is sometimes built in South America; the Villaneuva slab in Caracas, the December Apartments (Figure 17-9), is one example.

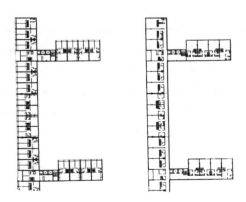

Figure 17-8 El Paraiso apartments, Caracas. Carlos Villaneuva, 1956.

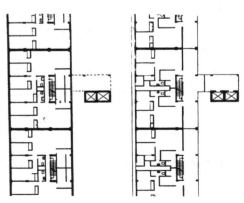

Figure 17-9 December Apartments, Caracas. Carlos Villaneuva, 1956.

Words and phrases

1. evolve   *v.*   发展，进展，进化
2. alternate-level   楼层变化的
3. evolutionary   *adj.*   进化的
4. extravagant   *adj.*   奢侈的，浪费的，过分的
5. double-height   两层高的

Section 2
Extensive Reading

Words and Buildings: A Vocabulary of Modern Architecture
Adrian Forty

The development of space as an architectural category took place in Germany, and it is to German writers that one must turn for its origins, and purposed. This immediately presents a problem for an English-language discussion of the subject, for the German word for space, Raum, at once signifies both a material enclosure, a "room", and a philosophical concept. As Peter Collins pointed out, "it required no great power of the imagination for a German to think of room as simply a small portion of limitless space, for it was virtually impossible for him to do otherwise" (1965, 286). In neither English nor French can a material enclosure so easily be linked to a philosophical construct, and consequently "space", as a translation for the German Raum, lacks the suggestiveness of the original. An example of the possibilities present in German, but lacking in English, can be seen in the translation of Rudolf Schindler's 1913 "Manifesto", discussed below.

As well as an awareness of the effects of translation upon the meaning of the term, we should also take into account the effects of time. "Space's" meanings in architecture are not fixed; they change according to circumstances and the tasks entrusted to it. When Denari and Lasdun enthuse about space, we should not assume that they mean by it what, say, Mies van der Robe meant by it in 1930. We must, as always, proceed by asking to what the category spoken of——in this case "space"——is being opposed: the reasons for valuing "space" in the 1990s are not the same as those in 1930. Despite the tendency of speakers to imply that they are talking about an immutable absolute, "space" is no less transient a term than any other in architecture.

Resources for Reference

http://en.wikipedia.org
http://www.amazon.com/
Adrian Forty. *Words and Buildings: A Vocabulary of Modern Architecture*. 2004: Thames & Hudson

Section 3
Tips for Translation

Translating long sentences 长句的翻译

英语中经常出现语法结构复杂的长句，句中包含较多的从句和修饰成分。特别是在科技英语等文体正式的文章中，长句较为多见。在长句中各种句子成分以关联词组成复杂的句子。相对而言，汉语句子一般较短，结构较为简单。因此，将英语中的长句翻译成汉语，是有一定的难度的。一般来说，既可以按照英语原文的结构顺序翻译，也可以按照汉语的逻辑顺序翻译。关键还是力求文字和意义的准确、连贯。

1. It is notable that the eight "treelike" plans he singles out for attack in the following selection represent a diverse set of the most respected and famous twentieth-century plans from Le Corbusier's plan for the new town of Chandigarh, India, based on his principles for a contemporary city, to Paolo Soleri's visionary megastructure of Mesa City in the Arizona desert.
值得注意的是，他（亚历山大）在下文中选出来予以批判的8个"树状"规划代表了20世纪最受尊重也最为著名一系列规划方案，包括柯布西耶基于当代城市规划原则的印度昌迪加尔新城规划，以及保罗·索拉里在亚利桑那沙漠中的梅萨城梦想城市。
在本句中，既有限制性的定语从句，也有过去分词短语作定语。

2. Many design professionals admire and attempt to incorporate into their designs for the built environment elements reflecting underlying human psychological and spiritual needs and cultural values.
不少专业设计人员赞赏并努力尝试把那些能够反映内在的人类心理和精神需求以及文化价值观的要素同建成环境的设计结合起来的做法。
在本句中"elements"是"incorporate into"的宾语，"reflecting…cultural values"是用作定语的分词短语，用于修饰"elements"。

3. The centre will cover a site of 975 873m^2 and includes new and redevelopment of existing facilities which at present consist of match fields for hockey, baseball, handball, shooting, archery and an equestrian centre, into a comprehensive international standard centre offering a main stadium seating 80000 spectators, cycling tacks, a natatorium, tennis centre, sports-themed sculpture park, news centre, hi-tech exchange centre, athletes centre etc.
中心将占地975873平方米，包括新建项目和对曲棍球、垒球、手球、射击、射箭和马术中心等现有设施的更新，从而建成一个具有国际标准的综合性体育中心，包括能容纳8万名观众的主体育场、自行车赛场、游泳馆、网球中心、体育主题雕塑公园、新闻中心、高技术信息中心和运动员中心等。

本句虽然较长，但是结构相对简单，因此在翻译时可以按照英文的结构顺序。

4. Actually barrel roofs of reinforced concrete in the shape of half-cylinders with curvatures down are commonly and inexpensively used in industrial buildings, since they can be poured on the same cylindrical formwork, which can be moved from one location to another and reused to pour a large number of barrels on the same form.

事实上向下弯曲的半圆柱状钢筋混凝土桶形结构的屋面是工业建筑的常用和经济的结构方式，因为它们可以采用同样的圆柱形模子进行浇筑，这些模子搬运方便，在浇筑同样形状的桶形构件时可以重复使用。

在本句翻译中，根据句子的内在逻辑关系，添加一些连接词，使译文顺畅。

Section 4
Tips for Writing

Keep Sentences short

It's easier to make a point clearly if you try not to exceed 18 words per sentences. Avoid cramming in too many ideas——one idea per sentence is plenty.

Read the following example from a proposal.

Our multi-disciplinary team offers not only capabilities in space programming, site planning, architectural design, structural, mechanical, and electrical engineering, but also provides services in the areas of financial feasibility studies and environmental assessment, as well as in the administration of the construction contract and in the development of post-occupancy monitoring systems, all of which are critical elements in the successful implementation of a viable construction program.

The meaning is hidden in a jungle of verbiage. Try to pinpoint the ideas, and make each one into a separate sentence. One solution:

Our team offers services in several categories:
- Space programming, site planning, architectural design including construction contract administration, as well as structural, mechanical, and electrical engineering.
- Financial feasibility studies and environmental assessments.
- Construction contract administration and development of post-occupancy monitoring systems.

Unit 18

Section 1
Intensive Reading

Modern Housing Prototypes
Roger Sherwood

Part IX

Double-Loaded Corridor Systems

Double-loaded corridor slabs are more numerous than the single-loaded type, and a greater variety of types are possible. Able to accommodate either single-orientation units (corridor every floor) or double-orientation units (skip-stop), this building type has much greater flexibility than single-loaded buildings. Le Corbusier's Unité d'sHabitation at Marseilles of 1952 (Figure 14 – 14) popularized the double-loaded, skip-stop section, and it appears frequently thereafter in many countries.

Double-loaded system; corridor every floor (Figure 18 – 1).
Double-loaded slabs with a corridor every floor are especially sophisticated, popular, and practical in the United States, where fire codes until recently rendered skip-stop systems virtually impossible. This type of building is Mies van der Rohe's stock-in-trade. His Lake Shore apartments in Chicago of 1948 (Figure 12 – 3) and the apartments in Baltimore (Figure 12 – 2) are typical and set the pattern for much that was to follow——not only in the organization but also in the image of the expensive, glass-walled residential skyscraper. Although not as popular in Europe, similar types such as the Nytorp slab in Malmo by Jaenecke and Samuelson (Figure 18 – 7) do on occasion appear.

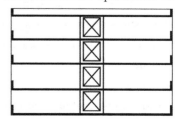

Figure 18 – 1

Double-loaded system; corridor every second floor. (Figure 18–2)
By far the more common double-loaded types follow the Marseilles Unité example, with corridors every second or third floor. The Lincoln Estate slab by Martin of 1960 in London (Figure 11–4) uses a system of corridors every other floor and an interlocking system of two-level units with living room at corridor level and internal stairs to bedrooms above on the opposite side of the building.

Double-loaded system; corridor every third floor (Figure 18–3).
Le Corbusier's section (Figure 18–8), with corridors every third floor, also uses a system of interlocking units. Unlike Lincoln Estate, however, the living room has a two-story volume and the bedrooms above run through the building. Entrance to one unit is at the living room level and in the other at the balcony level, with the double-orientation part of the apartment below. This is a much-copied scheme; other variations include the Neuwil block (Figure 18–9) by the Metron group of 1962 (although the units here do not interlock) and the Angerer slab at Munich of 1960 (Figure 18–10), a similar type with entrance off the corridor to one unit and stairs to units above and below. Each apartment here, like the Metron slab, is only one floor high. Sert follows this pattern in Peabody Terrace, the married students' housing at Harvard (Figure 14–11), one of the few alternate-level corridor buildings built in the United States until very recently.

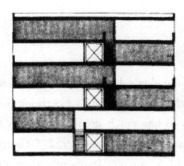

Figure 18–2

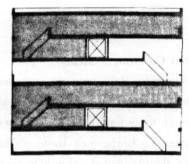

Figure 18–3

Double-Loaded Split-Level Systems

A final variation of the double-loaded corridor system is the split-level type. It comes with corridors every second and third floor or with the corridor in alternating positions in the slab. The idea of the split-level scheme is that one has to climb stairs up or down only one-half level from the corridor. Generally, both single- orientation and double-orientation units are used to get a mix of large and small apartments. The smaller units are usually single-loaded along one side of the corridor while the larger are split-level, usually with sleeping spaces on one side and living area on the other for a double-

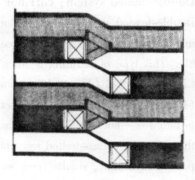

Figure 18–4

orientation, dumbbell type.

Double-loaded split-level system; corridor every second floor, alternating position. (**Figure 18–4**)

An example of this type is the apartment house in Germany by Schmiedel of 1960 (Figure 15–3), where the corridor is always double-loaded but asymmetrically positioned in section in alternating fashion.

Double-loaded split-level system; corridor every third floor. (**Figure 18–5**)

The Ramat Hadar slab at Haifa by Mansfeld and Calderon (Figure 18–11) is the example of this split-level arrangement, with the corridor always occurring in the same position in section.

Double-loaded split-level system; corridor every third floor, alternating position. (**Figure 18–6**)

This type became well-known from the van den Broek and Bakema tower at the Hansa project in Berlin in 1956 (Figure 15–4). Villaneuva, however, was proposing the same system at about the same time for a slab project in Caracas (Figure 18–12). The split-level types not only produce very compact buildings with few corridors and minimum walk up or down to each apartment but also create some spatial expansion within the unit because one can see up or down the stairs into opposite halves of the apartment, giving the impression of one large space. The alternating-position corridor scheme also gives larger spaces on one side at each level, thereby accommodating the need to have larger living spaces as well as a mix of unit sizes.

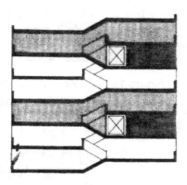   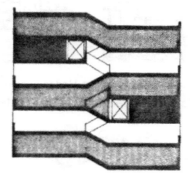

Figure 18–5   Figure 18–6

There are countless variations of each typical section. Some buildings would seem to escape classification at all, such as the amorphous group by Bruyère in Sausset-les-Pins of 1964 (Figure 18–13) or Habitat by Moshe Safdie in Montreal of 1964 (Figure 18–14), which, in its built form, does not seem to exhibit any consistent notion about the combination of units. Still other projects, for

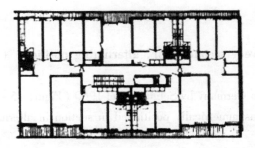

Figure 18-7 Nytorp apartments, Malmo, Sweden. Jaenecke and Samuelson, 1959.

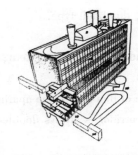

Figure 18-8 unité d'Habitation, Marseilles. Le Corbusier, 1952.

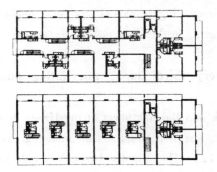

Figure 18-9 Apartment block, Neuwil-Wohlen, Switzerland. Metron, 1962.

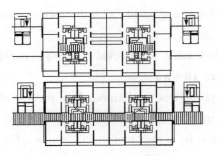

Figure 18-10 Apartment block, Munich. Fred Angerer, 1960.

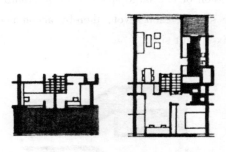

Figure 18-11 Ramat Hadar apartments, Haifa. Mansfeld and Calderson, 1964.

Figure 18-12 Apartment slab, Caracas. Carlos Villaneuva, 1956.

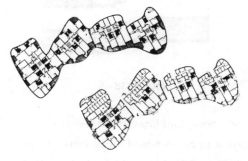

Figure 18-13 Apartments, Sausset-les-Pins, France. André Bruyere, 1964.

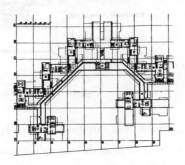

Figure 18-14 Habitat, Montreal. Moshe Safdie, 1964.

instance the Ziggurat in Israel by Gerstel of 1964 (Figure 18-15), do not seem to fit into any building category. Some examples seem bizarre but are really just permutations of stock types. Le Corbusier's Durand project in Algiers of 1933 (Figure 18-16), a strange cantilevered step-section building, is really just a double-loaded corridor, skip-stop type in which units cantilever and diminish in size toward the top of the building; Aalto's tower at Bremen (Figure 12-6), which seems quite unconventional, is just a simple single-loaded corridor plan.

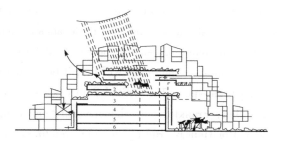

Figure 18-15 Ziggurat, Israel. Leopold Gerstel, 1964.

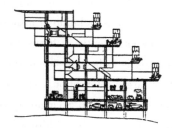

Figure 18-16 Durand apartments, Algiers. Le Corbusier, 1933.

Of the thirty-two case studies that follow, it is significant that only four are from the United States. Two of the four (Suntop and El Pueblo) are groups of semidetached houses of a low-density type, and only two (Peabody Terrace and Price Tower) are high-density projects. Among the more common varieties of urban housing——the rowhouse, party-wall building, blocks, and slabs——no American examples are included; representatives of most of these types can be found, but the choice is limited.

High-density housing in the United States has tended to be either luxury high-rise buildings or racially segregated low-income developments. The luxury housing is publicized and monumentalized (Mies van der Rohe's Lake Shore Drive apartments in Chicago, for example). But more typical has been the Bedford-Stuyvesant/Pruitt-Igoe kind of urban housing——anonymous, overcrowded, racially segregated, and economically depressed. It is doubtful if architecture can ever be the means to social deliverance——the problem is one of national attitudes and policies. Ironically, the dramatic, explosive demolition of the housing slabs in St. Louis (Figure 18-17) happened to

Figure 18-17 Destruction of the Pruitt-Igoe slabs, St. Louis, Missouri, 1972.

buildings which the inhabitants found well designed in some respects but which could not survive an extremely hostile socioeconomic environment. If the Pruitt-Igoe slabs had been built on the outskirts of almost any European city, they probably would have provided useful and acceptable housing.

Americans, with a continent of land available to them, have traditionally taken detached housing as a norm, and until recently a majority of middleclass families have been able to afford it. From 1955 to 1975, however, housing costs rose at almost twice the rate of income;① this trend, and the pressures of population growth and fuel shortages, suggest that new housing in decades to come will be preponderantly in forms other than that of the suburban single-family home. If higher-density housing is to become the norm for middle-income families, Americans will find it beneficial to look to a larger international scene for useful housing prototypes. The United States has very few that can compare with Spangen Quarter, Siedlung Halen, Frankfurt, or Siemensstadt; and it has had no national housing exhibitions such as Weissenhof or Hansaviertel to which outstanding architects and planners have been invited.

The buildings that follow are presented as case studies of different types of housing from throughout the world, beginning with the lower-density building types and ending with the high-density types. Included are detached housing (excluding the detached single-family house), rowhouses, terrace houses, party-wall housing, large courtyard housing, slabs, and towers. Each project is described in terms of the history of its development and its importance as a housing prototype. They are intended only as a representative sampling; obviously by no means can all the pertinent housing prototypes be covered in thirty-two examples. These particular buildings were chosen because they represent well-known models of a particular housing type——the Unité d'Habitation of Le Corbusier or Siedlung Halen by Atelier 5 for instance——or because they are particularly revealing examples of a type, such as the Vienna Werkbund Exposition rowhouse of Lurcat or Michiel Brinkman's Spangen housing. All of them, in my judgment, reward study.

Words and phrases

1. flexibility   n.   弹性，适应性
2. popularized   adj.   普及的，大众化的
3. stock-in-trade   n.   存货，惯用手段
4. asymmetrically   adv.   不均匀地，不对称地
5. amorphous   adj.   无定形的，无组织的
6. bizarre   adj.   奇异的
7. cantilever   n.   悬臂

---

① From a report published by the National Association of Home Builders, Washington, D. C., 1975. The NAHB director of economics derived the contents of this report from statistics furnished by the Department of Housing and Urban Development.

8. racially   *adv.*   按人种
9. deliverance   *n.*   释放，意见，判决
10. preponderantly   *vi.*   占优势，超过，胜过
11. pertinent   *adj.*   有关的，中肯的

Section 2
Extensive Reading

Pattern Language
Christopher Alexander

Staircase as a Stage
A staircase is not just a way of getting from one floor to another. The stair is itself a space, a volume, a part of the building; and unless this space is made to live, it will be a dead spot, and work to disconnect the building and to tear its processes apart.

Figure 18-18

Our feelings for the general shape of the stair are based on this conjecture: changes of level play a crucial role at many moments during social gatherings; they provide special places to sit, a place where someone can make a graceful or dramatic entrance, a place from which to speak, a place from which to look at other people while also being seen, a place which increases face to face contact when many people are together.

If this is so, then the stair is one of the few places in a building which is capable of providing for this requirement, since it is almost the only place in a building where a transition between levels occurs naturally.

This suggests that the stair always be made rather open to the room below it, embracing the room, coming down around theouter perimeter of the room, so that the stairs together with the room form a socially connected space. Stairs that are enclosed in stairwells or stairs that are free standing and chop up the space below, do not have this character at all. But straight stairs, stairs that follow the contour of the walls below, or stairs that double back can all be made to work this way.

Furthermore, the first four or five steps are the places where people are most likely to sit if the stair is working well. To support this fact, make the bottom of the staircase flare out, widen the steps, and make them comfortable to sit on.

Finally, we must decide where to place the stair. On one hand, of course, the stair is the key to movement in a building. It must therefore be visible from the front door; and, in a building with many different rooms upstairs, it must be in a position which commands as many of these rooms as possible, so that it forms a kind of axis people can keep clearly in their minds.

However, if the stair is too near the door, it will be so public that its position will undermine the vital social character we have described. Instead, we suggest that the stair be clear, and central, yes——but in the common area of the building, a little further back from the front door than usual. Not usually in the entrance room, but in the COMMON AREA AT THE HEART. Then it will be clear and visible, and also keep its necessary social character.

Therefore:

Place the main stair in a key position, central and visible. Treat the whole staircase as a room (or if it is outside, as a courtyard.) Arrange it so that the stair and the room are one, with the stair coming down around one or two walls of the room. Flare out the bottom of the stair with open windows or balustrades and with wide steps so that the people coming down the stair become part of the action in the room while they are on the stair, and so that people below will naturally use the stair for seats.

Resources for Reference

http://www.amazon.com
http://en.wikipedia.org
http://www.patternlanguage.com
Christopher Alexander. *Pattern Language*. 1977: Oxford University Press

Section 3
Tips for Translation

Translating meaning in an appropriate manner 符合汉语习惯的翻译

1. The remnants of Pompeii are enough to *suggest* the splendor of Roman empire thousands of years ago.

庞贝的遗址足以让人想象几千年前罗马帝国的辉煌。

2. *It suddenly occurred to me* that there would be lecture this afternoon.
我突然想起来下午有个讲座。

3. On one hand, curved surfaces are believed to be more complex to design than the flat rectangular shapes we are so used to. Usually quite the opposite is true.
一方面，人们认为相对于我们熟悉的平面矩形结构，曲面结构在设计上更为复杂。通常情况恰恰相反。

4. The engineer would only allow the roof to be built if two concrete arches were erected between its diagonally opposite comers "to support it."
只有当两个混凝土拱券在对角方向被支撑起来的时候，结构工程师才认为屋顶的建造是可行的。

　　在上面例句的翻译中，如果"suggest"、"It suddenly occurred to me"、"quite the opposite is true"和"would only allow the roof to be built"，如果一定要找到对等的汉语表达方式进行翻译，或者按照字面意思直译，会具有相当大的困难，并影响翻译的畅达。在这种情况下，应当充分认识到英语和汉语在表达方式上的区别，注重意义的传达，进行合乎汉语语言习惯的翻译。

Section 4
Listening Practice

Please watch the video and answer the questions below.

1. Describe the envolving process of the prototype model for the IN Project.
2. In Steven Holl's opinion, the project architecturally stands as a manifesto for ecological development. Why?
3. How do the spheres relate to each other?

Words and Phrases:

1. neuroscience   *n.*   神经系统科学
2. Venn diagram   文氏图，用于显示元素集合重叠区域的图示
3. tesseract   *n.*   立方体的四维模拟，超正方体
4. guesthouse   *n.*   宾馆
5. rock outcroppings   岩石露头

6. slate v. 用石板瓦盖，用板岩覆盖（例如屋顶）；指定
7. inspiration n. 灵感
8. manifesto n. 宣言
9. ecological adj. 生态学的

Resource for Reference

https://www.youtube.com/watch?v=OmjrTghS6gw

Unit 19

Section 1
Intensive Reading

Form-Resistant Structures
Mario Salvadori

Part I

Mario G. Salvadori, the renowned Columbia professor who worked to link the fields of structural engineering and architecture and served as a consultant on the Manhattan Project, died of natural causes on June 25 at Mt. Sinai Hospital. He was 90 (Figure 19-1).

Salvadori was the author of ten books on architectural structures (including Structural Design in Architecture, 1967) and five books on applied mathematics (including Numerical Methods in Engineering, 1953). He had taught at Columbia since 1940, and, at the time of his death, was the James Renwick Professor Emeritus of Civil Engineering and Applied Science and Professor of Architecture Emeritus.

Figure 19-1

"Aside from being a brilliant mathematician and an outstanding engineer whose office was considered an ideal training ground for young engineers, Mario Salvadori was also a charismatic teacher of structures at the School of Architecture, Planning and Preservation at Columbia University," said Kenneth Frampton, Ware Professor of Architecture at Columbia. "With his boundless engineering knowledge and deep sense of public commitment, he made a unique and wide-ranging contribution to both the University and to society at large. He will be greatly missed."

Salvadori investigates building structure as an integral part of architectural form and the design process. Much of what must be learned about structures by architects can become mired in abstract mathematical analyses and formulae. Salvadori elevates this technical dimension of the architectural process to show how structural form can and should be inseparable from the more subjective facets of design. In this article from his book Why Buildings Stand Up, Salvadori demonstrates how geometric shapes not only signify inherent structural principles but also can be used as powerful design elements to enhance meaning and significance in architectural form. Through the use of flat and folded plates, curvilinear forms, vaults, and interlocking building components, he provides a narrative that creates an architectural dialogue through the medium of a structural language. This article shows how the apparently conflicting intentions of the architect and the engineer can be fused into a unified creative process.

Words and phrases

1. resistant *adj.*   抵抗的，反抗的，耐久的
2. renowned *adj.*   著名的，有声誉的
3. emeritus *adj.*   名誉退休的
4. charismatic *adj.*   有魅力的
5. commitment *n.*   承担义务
6. at large   普遍地
7. integral *adj.*   完整的，整体的
8. be mired in   陷入
9. subjective *adj.*   主观的/objective 客观的
10. geometric *adj.*   几何的，几何学的
11. signify *v.*   表示，意味
12. inherent *adj.*   固有的，内在的
13. plate *n.*   板
14. curvilinear *adj.*   曲线的
15. vault *n.*   拱顶
16. interlocking *adj.*   联锁的
17. building component   建筑构件
18. narrative *n.*   叙述
19. architectural dialogue   建筑对话
20. be fused into   （被）融入……

Section 2
Extensive Reading

Kenneth Frampton

Kenneth Frampton (born 1930, Woking, UK) is a British architect, critic, historian and Professor of Architecture at the Graduate School of Architecture and Planning, Columbia University, New York.

Frampton studied architecture at Guildford School of Art and the Architectural Association School of Architecture, London. Subsequently he worked in Israel, with Middlesex County Council and Douglas Stephen and Partners (1961 ~ 1966), during which time he was also a visiting tutor at the Royal College of Art (1961 ~ 1964), tutor at the Architectural Association (1961 ~ 1963) and Technical Editor of the journal Architectural Design (AD) (1962 ~ 1965).

Frampton has also taught at Princeton University (1966 ~ 1971) and the Bartlett School of Architecture, London (1980). He has been a member of the faculty at Columbia University since 1972, and that same year he became a fellow of the Institute for Architecture and Urban Studies in New York (whose members also included Peter Eisenman, Manfredo Tafuri and Rem Koolhaas) and a co-founding editor of its magazine Oppositions.

Frampton is well known for his writing on twentieth-century architecture. His books include *Modern Architecture: A Critical History* (1980; revised 1985 and 1992) *and Studies in Tectonic Culture* (1997). Frampton achieved great prominence (and influence) in architectural education with his essay "Towards a Critical Regionalism" (1983) though the term had already been coined by Alex Tzonis and Liliane Lefaivre. Also, Frampton's essay was included in a book *The Anti-Aesthetic*. Essays on *Postmodern Culture*, edited by Hal Foster, though Frampton is critical of postmodernism. Frampton's own position attempts to defend a version of modernism that looks to either critical regionalism or a "momentary" understanding of the autonomy of architectural practice in terms of its own concerns with form and tectonics which cannot be reduced to economics (whilst conversely retaining a Leftist viewpoint regarding the social responsibility of architecture).

In 2002 a collection of Frampton's writings over a period of 35 years was collated and published under the title Labour, Work and Architecture.

Select list of Frampton's writings
1. Studies in Tectonic Culture: The Poetics of Construction in Nineteenth and Twentieth Century Architecture. MIT Press, Cambridge, Mass. , 2001.
2. Modern Architecture: A Critical History (World of Art), Thames & Hudson, London, Third edition (1992).
3. Le Corbusier (World of Art). Thames & Hudson, London, 2001.
4. Labour, Work and Architecture. Phaidon Press, London, 2002.
5. The Evolution of 20th-Century Architecture: A Synoptic Account. Springer, New York, 2006.

Resource for Reference

http://eng.archinform.net

Section 3
Tips for Translation

Extension 词义引申

在翻译中为了选择准确的用词和表达方式，不必完全拘泥于英文原文，在有的时要对原文中的某些成分进行意义上的引申，反而能够更加准确地表达原文的意思。

1. One of the most commonly *encountered* combinations of cylindrical surfaces is the *groined vault* of the Gothic cathedrals.
最常见的柱状曲面的组合就是哥特式教堂的交叉拱。
"encounter"的原意是"遭遇"、"碰到"的意思，在本句中可以结合"commonly"译为"常见"。

2. The largest roof of this kind has been *erected* in Denver, Colorado, and measures 112 feet by 132 feet, with a three-inch thickness.
此类屋顶中最大的建成科罗拉多的丹佛，有112英尺乘以132英尺大小，厚度3英寸。
"erect"的本意是"竖立"的意思，直接用于汉译较为生硬，在这里可以引申为"建成"。

3. The mode of support of a barrel influences its load-carrying *action*.
桶形结构的支承方式会影响其承载效果。
"action"不能译为"行为"，引申为"效果"较为妥当。

4. If a barrel is supported all along its two longitudinal edges it acts as a series of arches built one next to the other and develops out-pushing thrusts, which must be *absorbed* by buttresses or tie-rods as in any arch.
如果支承位于桶形结构的长向，它就表现为一系列的拱券，并产生向外的推力。这种推力必须靠扶壁或系杆来抵消，对于任何拱券都是这样。
"absorb"本来是"吸收"的意思，用在此句的翻译中欠妥当，不如引申为"抵消"或者"平衡"。

5. To *visualize* how a curved surface can be obtained by means of straight lines, connect by inclined straight-line segments the points of two equal circles set one above the other.

为了形象地说明如何通过直线来获得曲面，可以用倾斜的直线段把一上一下的同样两个圆上的点连起来。

"visualize"本意是"形象化"、"视觉化"的意思，这里可以结合上下文翻译为"形象地说明"。

Unit 20

Section 1
Intensive Reading

Form-Resistant Structures
Mario Salvadori

Part II

GRIDS AND FLAT SLABS

Ever since the beginning of recorded history, and we may assume even earlier, people have gathered in large numbers for a variety of purposes be they religious, political, artistic, or competitive. The large roof, unsupported except at its boundary, arose to shelter these gatherings, evolving eventually into the huge assembly hall we know today.

As we shall see, no large roof can be built by means of natural or man-made compressive materials without giving the roof a curved shape, and this is why domes were used before any other type of cover to achieve large enclosed spaces. Even wood, a material that can span relatively short horizontal distances by beam action..., has to be combined in conical, cylindrical or spherical shapes whenever large distances are to be spanned.

Only after the invention of inexpensive methods of steel manufacture and the recent development of reinforced concrete did large flat roofs become possible. They have obvious advantages over dome roofs: their erection is simpler, and they do not waste the upper part of the space defined by the dome which is often superfluous, unnecessarily heated or air-conditioned.

The simplest structural system for a flat rectangular roof consists of a series of parallel beams supporting some kind of roofing material. But if all four sides of the rectangle to be covered can be used to support the roof beams, it becomes more practical to set the beams in two directions, at right angles to each other, thus creating a grid. This two-way system pays only if the two dimensions of the rectangle are more or less equal. Loads tend to move to their support through the shortest possible path and if one dimension of the roof is much larger than the other, most of the load will be carried by the shorter beams, even if the beams are set in a grid pattern.

A grid is a "democratic" structural system: if a load acts on one of its beams, the beam deflects, but in so doing carries down with it all the beams of the grid around it, thus involving the carrying capacity of a number of adjoining beams. It is interesting to realize that the spreading of the load occurs in two ways: the beams parallel to the loaded beam bend together with it, but the beams at right angles to it are also compelled to twist in order to follow the deflection of the loaded beam (Figure 20-1). We thus find that in a rectangular grid loads are carried to the supports not only by beam action (bending and shear) in two directions but by an additional twisting mechanism which makes the entire system stiffer. To obtain this twisting interaction the beams of the two perpendicular systems must be rigidly connected at their intersection, something which is inherent in the monolithic nature of reinforced concrete grids and in the bolted or welded connections of steel grids. Even primitive people know how to obtain such twisting action by interweaving the beams of their roofs so that any displacement of one beam entails the bending and twisting displacement of all the others (Figure 20-2).

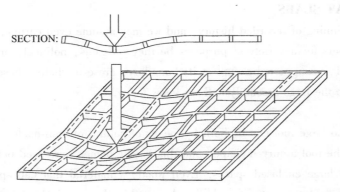

Figure 20-1　A rectangular grid of beams. [Saralinda Hooker and Christopher Ragus]

Though rectangular grids are the most commonly used, skew grids (Figure 20-3) have, beside aesthetic qualities, the structural and economic advantage of using equal length beams even when the dimensions of the grid are substantially different, thus distributing more evenly the carrying action between all the beams.

We have seen how grids of trusses rather than beams become necessary when spans are hundreds of feet long, and how space frames constitute some of the largest horizontal roofs erected so far,

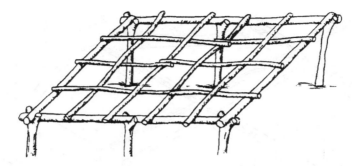

Figure 20-2   A woven grid of beams. [Saralinda Hooker and Christopher Ragus]

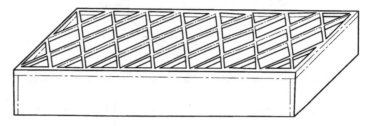

Figure 20-3   A skew grid. [Saralinda Hooker and Christopher Ragus]

covering four or more acres without intermediate supports. We must now go one step back to discover how an extension of the grid concept has become the principle on which most of the floors and roofs of modern buildings are built.

Let us imagine that the beams of a rectangular grid are set nearer and nearer to each other and glued along their adjacent vertical sides until they constitute a continuous surface. Such a continuous surface, called a plate or slab, presents all the advantages of a grid in addition to the ease with which it can be poured on a simple horizontal scaffold when made out of concrete. Reinforced concrete horizontal slabs are the most commonly used floor and roof surfaces in buildings with both steel and concrete frames all over the world. Their smooth underside permits a number of things to hang——pipes and ducts, for instance——without having to go around beams. The setting of the slab reinforcement on flat wooden scaffolds makes the placing of the steel bars simple and economical. In European countries concrete slabs are sometimes made lighter by incorporating hollow tiles (Figure 20-4). Through the strength of their burnt clay these tiles participate in the slab structural action, which is the same in all slabs whatever their material.

Actually slabs, besides carrying loads by bending and twisting like grids of beams, have an additional capacity which makes them even stiffer and stronger than grids. This easily understood capacity derives from the continuity of their surface. If we press on a curved sheet of material attempting to flatten it, depending on its shape, the sheet will flatten by itself or have to be stretched or sliced before it can be made flat. For example, a sheet of paper bent into a half-cylinder and

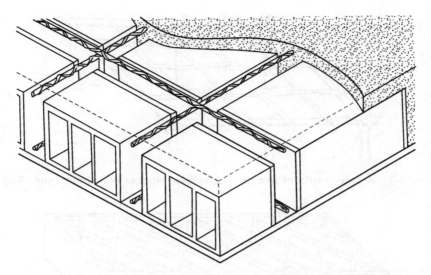

Figure 20-4 Tile-concrete floor slab. (Saralinda Hooker and Christopher Ragus)

then released flattens by itself [Figure 20-5 (a)]. It is said to be a developable surface (from the idea "to unfold" contained in the verb "to develop"). But if we cut a rubber ball in half, producing a small spherical dome, the dome will not flatten by itself if we lay it on a flat surface. Neither will it become flat if we push on it. It only flattens if we cut a large number of radial cuts in it or if, assuming it is very thin. it can be stretched into a flat surface [Figure 20-5 (b)]. The dome (and actually all other surfaces except the cylinder) are non-developable, unflattening surfaces. Because they are so hard to flatten, they are also much stiffer than developable surfaces. (It will be more obvious why non-developable surfaces are better suited to build large roofs once we learn how such roofs sustain loads.)

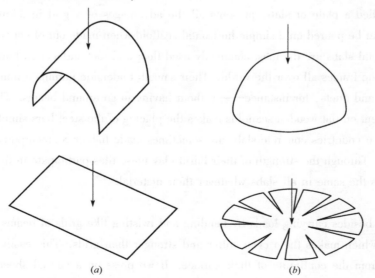

Figure 20-5 Developable (a) and non-developable (b) surfaces.
(Saralinda Hooker and Christopher Ragu)

Returning now to the behavior of a flat slab, we notice that under load it becomes "dished" —— it acquires the shape of a curved surface, with an upward curvature (Figure 20-6). If it is supported only on two opposite parallel sides, it becomes a slightly curved upside-down cylinder, but if it is supported on four sides, or in any other manner, it acquires a non-developable shape. Just as the half-ball had to be stretched to be changed from a dome to a flat surface, the plate has to be stretched to change it from a flat to a dished surface. Hence the loads on it, besides bending it and twisting it, must stretch it, and this unavoidable stretching makes the slab even stiffer. Therefore we should not be amazed to learn that plates or slabs can be made thinner than beams. While a beam spanning twenty feet must have a depth of about one-and-half feet, whether it is made of steel, concrete or wood, a concrete slab covering a room twenty-feet square can be made one foot deep or less.

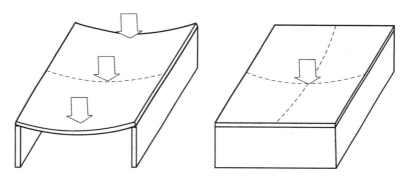

Figure 20-6 Flat slabs dished by loads. (Saralinda Hooker and Christopher Ragus)

Words and phrases

1. compressive   *adj.*   压缩的
2. enclose   *v.*   围护/enclosure
3. span   *n.*   跨度/*v.* 跨越
4. conical   *adj.*   圆锥的，圆锥形的
5. cylindrical   *adj.*   圆柱的，圆柱形的
6. spherical   *adj.*   球的，球形的
7. reinforced concrete   钢筋混凝土
8. superfluous   *adj.*   多余的
9. roofingmaterial   屋面材料
10. load   *n.*   荷载
11. deflect   *v.*   （使）偏斜，（使）偏转
12. carrying capacity   承载力
13. bend   *v.*   弯曲
14. shear   *v.*   剪切
15. twisting mechanism   扭转机制

16. perpendicular *adj.* 垂直的/vertical *adj.* 竖直的
17. monolithic *adj.* 整体的
18. bolt *n.* 螺栓/*v.* 拴接
19. weld *n.* 焊缝/*v.* 焊接
20. displacement 位移
21. skew *adj.* 歪斜的
22. substantially *adv.* 充分地，实质上地
23. distribute *v.* 分布，分发
24. truss *n.* 构架，桁架，钢梁
25. intermediate *n.* 媒介，中介
26. constitute *v.* 组成
27. frame *n.* 结构框架
28. pipes and ducts 管线
29. scaffold *n.* 脚手架
30. tile *n.* 面砖
31. hollow tile 空心砖
32. burnt clay 黏土砖
33. stiff *adj.* 有刚度的/strong *adj.* 有强度的
34. developable surface 可展曲面/non-developable surface 不可展曲面
35. curvature *n.* 弯曲
36. upside-down *adj.* 颠倒的

Section 2
Extensive Reading

Pier Luigi Nervi

Pier Luigi Nervi (June 21, 1891 ~ January 9, 1979) was an Italian architect and engineer. He studied at the University of Bologna and qualified in 1913. He is renowned for his brilliance as a structural engineer and his novel use of reinforced concrete.

Pier Luigi Nervi was born in Sondrio and attended the Civil Engineering School of Bologna, from which he graduated in 1913. After graduation, Nervi joined the Society for Concrete Construction. Nervi spent several years in the Italian army during World War I from 1915 to 1918, when he served in the Corps of Engineering. His formal education was quite similar to that experienced by today's civil engineering student in Italy.

Nervi began practicing civil engineering after 1923, and built several airplane hangars amongst his contracts. During 1940s he developed ideas for a reinforced concrete which helped in the rebuilding of many buildings and factories throughout Western Europe, and even designed/created a boat hull that was comprised of reinforced concrete as a promotion for the Italian government.

Nervi also stressed that intuition should be used as much as mathematics in design, especially with thin shelled structures. He borrowed from bothRoman and Renaissance architecture to create aesthetically pleasing structures, yet applied structural aspects such as ribbing and vaulting often based on nature. This was to improve the structural strength and eliminate the need for columns. He succeeded in turning engineering into an art by taking simple geometry and using sophisticated prefabrication to find direct design solutions in his buildings.

Pier Luigi Nervi was educated and practised as a "building" engineer (ingeniere edile) in Italy, at the time (and to a lesser degree also today), a building engineer might also be considered an architect. After1932, his aesthetically pleasing designs were used for major projects. This was due to the booming number of construction projects at the time which used concrete and steel in Europe and the architecture aspect took a step back to the potential of engineering. Nervi successfully made reinforced concrete the main structural material of the day.

Most of his built structures are in his nativeItaly, but he also worked on projects abroad. Nervi's first project in the United States was the George Washington Bridge Bus Station. He designed the roof which consists of triangle pieces which were poured in place. This building is still used today by over 700 buses and their passengers. He was awarded Gold Medals by the AIA, RIBA and the Académie d'architecture and taught as a professor of engineering at Rome University from 1946 to 1961.

Resource for Reference

http://en.wikipedia.org/wiki/Pier_Luigi_Nervi#Biography

Section 3
Tips for Oral Presentation

Choose an appropriate method of speaking

There are five methods of speaking as shown in Figure 20-7. The method you choose will have some bearing on your ultimate success in reaching your objective.

The most formal way to deliver a talk is to *memorize* a manuscript from start to finish. This method has the advantage of ensuring that you'll say exactly what you wanted to say before

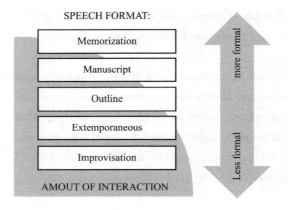

Figure 20-7

the presentation, but several disadvantages come to mind. Firstly, memorization is hard work for most people. An eight-minute speech requires four to five pages of double-spaced typing. Secondly, you are also possible to the risk of forgetting your lines. Thirdly, memorized speeches can sound canned unless you are gifted in the skill of bring emotional nuance to each idea you bring up.

It may be the single worst way to convey information orally to read from a ***manuscript***. In the real world, where designers practice, reading from a manuscript is a guaranteed ticket to a bored and indifferent audience, and to the almost certain failure of reading your objective. Its lone advantage is that you will say only the words you intended, but this advantage is trifling compared to the suffering you will inflict on your audience.

Speaking from an ***outline*** means to use brief fragments of ideas to jog your memory of what to speak about, using language that is fresh each time you say it. Of particular importance is that the language you use not be the exact wording of your outline, in which case you find yourself reading from a very brief, telegraphic, and at times, incomprehensible script. An outline can help keep you organized and on the subject, but without the deadly monotony of reading from a script.

Extemporaneous speaking is the one most likely to achieve your objective for your presentation. It requires a great deal of preparation and thorough knowledge of the subject at hand. In that way, it is like memorization, with one crucial difference: in memorization, the focus is on the particular words to be conveyed, on preserving exactly their order and structure as it was first written down. In extemporaneous speaking, the focus is not on the particular words, but on the ideas that are to be conveyed. It is in speaking extemporaneously about a subject, with obvious passion, keen interest, and deep experience, that a speaker can most easily and effectively persuade his or her audience to assent to his or her primary objective.

Improvisation means making it up as you go along. Potentially, it can be a masterful work of performance art. At its worst, however, it makes you look unprepared, lackadaisical, disinterested, and unprofessional. While this technique is daring and can be spectacular if done well, the risk of failure is too great to recommend it for design professionals.

Resource for Reference

David Greusel, AIA. 2002. Architect's Essentials of Presentation Skills, John Wiley & Sons, Inc., New York.

Unit 21

Section 1
Intensive Reading

Form-Resistant Structures
Mario Salvadori

Part III

When slabs have to span more than fifteen or twenty feet, it becomes economical to stiffen them on their underside with ribs, which can be oriented in a variety of ways. Nervi made use of Ferrocemento, a material he perfected, to build forms in which to pour slabs stiffened by curved ribs, which are oriented in the most logical directions to transfer the loads from the slab to the columns. These curved ribs, moreover, give great beauty to the underside of the slabs (Figure 21-1). Ferrocemento is a material consisting of a number of layers of welded mesh set at random, one on top of the other, and permeated with a concrete mortar, a mixture of sand, cement, and water (Figure 21-2). Flat or curved elements of Ferrocemento can be built only one or two inches thick, with exceptional tensile and compressive strength due to the spreading of the tensile steel mesh through the high-strength compressive mortar. First used only as a material to build complex molds in which to pour reinforced concrete elements, it later was transformed by Nervi into a structural material itself. Some of the masterpieces of Nervi owe their extraordinary beauty and efficiency to the use of Ferrocemento.

Genius often consists of an ability to take the next step, and Nervi took it by realizing that Ferrocemento would be an ideal material for building boats. His lovely ketch Nennele (Figure 21-3) was the first, but a large number of sailing boats have been built, mostly in Australia and the United States, with Ferrocemento hulls. They are easy to manufacture and even easier to repair in case of an accident.

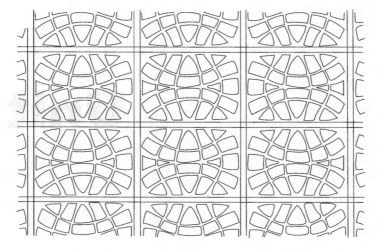

Figure 21-1　Nervi's slab with curved ribs.
(Saralinda Hooker and Christopher Ragusj)

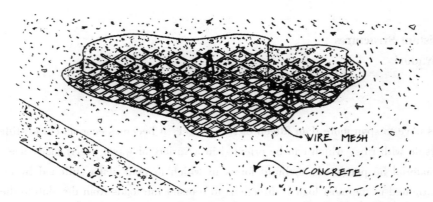

Figure 21-2　Ferrocemento mesh reinforcement.
(Saralinda Hooker and Christopher Ragus)

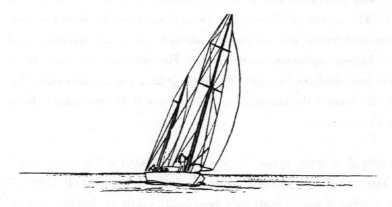

Figure 21-3　Nervi's ketch Nennele (of ferrocemento)
(Saralinda Hooker and Christopher Ragus)

STRENGTH THROUGH FORM

The stiffness of flat slabs, like that of beams, derives from their thickness: if too thin they become too flexible to be functional. It is one of the marvels of structural behavior that stiffness and strength of sheetlike elements can be obtained not only by increasing their thickness and hence the amount of required material, but by giving them curved shapes. Some of the largest, most exciting roofs owe their resistance exclusively to their shape. This is why they are called form-resistant structures.

If one holds a thin sheet of paper by one of its short sides, the sheet is incapable of supporting even its own weight-the paper droops down (Figure 21–4a). But if we give the side held a slight curvature up, the same sheet of paper becomes stiffer and capable of supporting as a cantilever beam not only its weight but also the small additional weight of a pencil or pen (Figure 21–4b). We have not strengthened the paper sheet by adding material to it; we have only curved it up. This principle of strength through curvature can be applied to thin sheets of reinforced concrete and has been efficiently used to build stadium roofs that may cantilever out thirty or more feet with a thickness of only a few inches (Figure 21–5). The shape of such roofs can be shown to be non-developable and hence quite rigid, but even developable surfaces, like cylinders, show enough strength (when correctly supported) to allow their use as structural elements. To demonstrate this property, try to span the distance between two books by means of a flat sheet of paper acting as a plate. The paper will sag, fold, and slide between the book supports. If instead the sheet of paper is curved up and prevented from spreading by the book covers, it will span the distance as an . arch (Figure 21–6). Again the curvature has given the thin paper its newly acquired stiffness and strength.

Nature knows well the principle of strength through curvature and uses it whenever possible to protect life with a minimum of material. The egg is a strong home for the developing chick, even though its shell weighs only a fraction of an ounce. The seashell protects the mollusk from its voracious enemy and can, in addition, sustain the pressure of deep water thanks to its curved surfaces. The same protection is given snails and turtles, tortoises and armadillos, from whom our medieval knights may have copied their curved and relatively light armor.

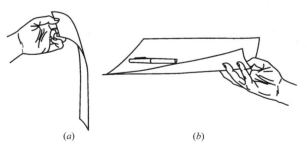

(a)   (b)

Figure 21–4 Paper sheet stiffened by curvature.
(Saralinda Hooker and Christopher Ragus)

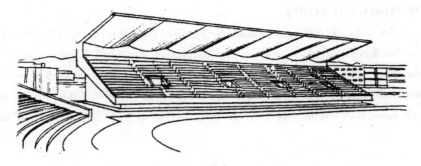

Figure 21-5 Stadium stands roof.
(Saralinda Hooker and Christopher Ragus)

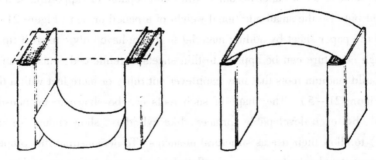

Figure 21-6 Paper sheet stiffened by cylindrical shape.
(Saralinda Hooker and Christopher Ragus)

Words and phrases

1. Ferrocemento *n.* 加筋水泥
2. welded mesh 焊接铁丝网
3. be permeated with 充满
4. mortar *n.* 灰浆
5. cement *n.* 水泥
6. tensile *adj.* 可拉长的，拉力的
7. ketch *n.* 双桅帆船
8. flexible *adj.* 柔韧的，有弹性的
9. marvel *n.* 奇迹
10. cantilever *n.* 悬臂梁/*v.* 悬挑
11. mollusk *n.* 软体动物
12. voracious *adj.* 贪婪的
13. armadillo *n.* 犰狳
14. medieval *adj.* 中世纪的
15. knight *n.* 骑士

16. armor   *n.*   盔甲，装甲

Section 2
Extensive Reading

Exhibition Building, Turin

The Creator's Words

"...two of my most interesting projects, the hangars built of pre-cast elements and the roof for the Turin Exhibition Halls, would have been impossible without a simultaneous invention of the structural method. They would have looked completely different if they had been built on the same principle but in a conventional technique."

——Pier Luigi Nervi from the Introduction of Jürgen Joedicke. The Works of Pier Luigi Nervi. pVII.

Details

"The hall is rectangular and covers an area of 240 feet × 309 feet. On one of the two shorter sides is a semi-circular apse. Windows are arranged in the corrugation of the prefabricated roof elements."

"A semi-circular apse 132 feet in diameter adjoins the main hall which is 240 feet long. Its roof consists of corrugated pre-cast units. The half-dome roof of the apse is also constructed with prefabricated elements."

"The vaulted construction of the hall consists of prefabricated elements which spring from in situ concrete abutments."

"The units are of 'ferro-cement' and have a length of approximately 15 feet and a width of 8 feet 3 inches. The thickness of the curved precast parts is less than 2 inches. This small thickness is achieved only by the increased rigidity through the corrugation and the transverse webs at either end. The individual units are joined by in situ concrete."
——Jürgen Joedicke. The Works of Pier Luigi Nervi. p59-62.

Resource for Reference

http://www.greatbuildings.com

Section 3
Tips for Translation

Translating words based on their exact meaning 翻译的准确性

在英译汉的过程中，应当根据句子本身和上下文，明确原文试图表达的含义，斟酌汉语的用词和表达方式，力求准确。要避免从词典上寻找汉译并不加甄别地直接套用。

1. He worked very hard during his life time to prove the real *nature* and potential reinforced concrete had in structural engineering.
他在有生之年致力于检验钢筋混凝土在结构工程上的特性和潜力。
"nature"既有"自然"的意思，也有"本质"、"特性"的意思。应当根据整句和上下文，明确词语的确切含义。

2. He also looked to solve problems by the simplest means possible.
他也希望通过最简单的方式来解决问题。
"look"在这里不是简单的"看"的意思，而是由"期待"的含义。

3. The mechanical and electrical *services* shall be environmental-friendly, reliable, practical, flexible, economical and intelligent.
机械和电力系统应当是环保、可靠、实用、弹性、经济和智能的。
本句中的"service"译为"系统"较符合汉语的相关表达习惯。

4. In the event of an outbreak of hostilities, the underground car parks, mechanical and *service* rooms will be converted into shelters for personnel and provisions.
在爆发战争的情形下，位于地下的停车、机械和后勤服务用房可以转变为人员和装备的掩蔽所。

与上句不同，本句中的"service"译成"后勤服务"较为妥当。

5. I admire her *ambition* to become a first-rate architect.
我赞赏她成为一流建筑师的志向。

6. The completion of Museo Guggenheim Bilbao greatly added to Frank Gehry's *reputation*.
比尔巴鄂古根海姆博物馆的落成使弗兰克·盖里声望大增。

7. Facing the radical social transformation made some people feel that idealism has gone out of life and that personal *ambition* and money have taken the place.
在巨大的社会变迁面前，有些人觉得生活中理想主义已经消失，取而代之的是个人的野心和金钱。

8. During the Second World War the Japanese soldiers won themselves a savage *reputation*.
二战期间，日本兵因残暴成性而臭名昭著。

比较一下5、6两句和7、8两句的翻译，同样的"ambition"和"reputation"，由于句子本身的感情色彩而采取不同的翻译，从而较为准确地表达原文的含义。

Unit 22

Section 1
Intensive Reading

Form-Resistant Structures
Mario Salvadori

Part IV

CURVED SURFACES

We owe to the greatest of all mathematicians, Karl F. Gauss (1777~1855), the discovery that all the infinitely varied curved surfaces we can ever find in nature or imagine belong to only three categories, which are domelike, cylinderlike, or saddlelike. *

How do the three categories differ? Consider the dome. Imagine cutting it in half vertically with a knife. The shape of the cut is curved downwards, and if you cut the dome in half in any direction, as you do when you cut a number of ice-cream-cake wedge slices, the shape of all the cuts is still curved downward (Figure 22 – 1). A domelike surface has downward curvatures in all its radial directions. By the way, if instead of cutting a dome we were to cut in half a soup bowl, we would find the shape of all the cuts to be curved up, whatever their radial direction. Domes and hanging Roofs, each with curvatures always in the same direction (either down or up), constitute the first of Gauss's categories. They are non-developable surfaces and have been used for centuries to cover large surfaces.

---

* Gauss was so great a man that he noted in a small book a number of discoveries "not worth publishing." When this booklet was found fifty years after his death, some of his "negligible" discoveries had been rediscovered and had made famous a number of his successors!

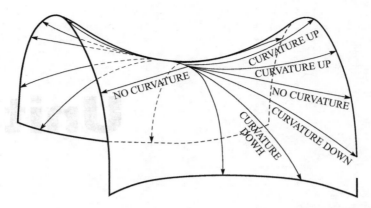

Figure 22-1　Vertical cuts in dome. (Saralinda Hooker and Christopher Ragus)

Let us jump to the third of Gauss's categories, the saddlelike surfaces. In a horse saddle the curve across the horse, defined by the rider's legs, is curved downward, but the curvature along the horse's spine, which prevents the rider from sliding forward or backward, is upward (Figure 22-2). Saddle surfaces are non-developable and are used as roofs because of their stiffness. The Spanish architect Felix Candela built as a saddle surface what is perhaps the thinnest concrete roof in the world. Covering the Cosmic Rays Laboratory in Mexico City, it is only half-an-inch thick.

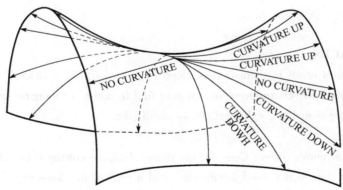

Figure 22-2　Vertical cuts in saddle. (Saralinda Hooker and Christopher Ragus)

Saddle surfaces have another property not immediately noticeable. As one rotates the saddle cuts from the direction across the horse to that along the horse, the curvature changes from down to up and, if one keeps going, it changes again from up to down. Therefore there are two directions along which the cuts are neither up nor down. They are not curved; they are straight lines (Figure 22-2). To prove this one has only to take a yardstick and place it across a saddle at its lowest point: the saddle is curved down, below the yardstick. If one then rotates the yardstick, keeping it horizontal, one finds that there is a direction along which the yardstick lies entirely on the surface of the saddle: in this direction the saddle has no curvature. Of course, if one rotates the yardstick in the opposite direction one locates the other no-curvature section of the saddle, which is symmetrical to the first with respect to the horse's axis. All saddle surfaces have two directions of no curvature. Cut along these directions, their boundaries are straight lines. This property makes the saddle shape an almost ideal surface with which to build roofs.

We can now go back to Gauss's second category, the cylinders. Imagine a pipe lying on the floor. If you cut vertically its top half——the half, say, with the shape of a tunnel——in any direction, you will notice that all of these cuts have a curvature down, except one: the cut along the pipe's axis is a straight line (Figure 22-3). The cylinder has no curvature in the direction of its axis. One may consider, then, the cylinder as a dividing line between the dome and the saddle. The saddles have two directions without curvatures, but as these two directions draw nearer and nearer, saddles become cylinders, with only one direction of no curvature. If this direction is now given a-down curvature, the cylinder becomes a dome. If instead of considering the upper part of a cylinder, we consider its lower half——the half, say, with the shape of a gutter——we find that the vertical sections of the gutter have curvatures up in all directions except one: the direction of the axis of the gutter. Hence, gutters and tunnels belong to the same category of surfaces having only one direction of no curvature.

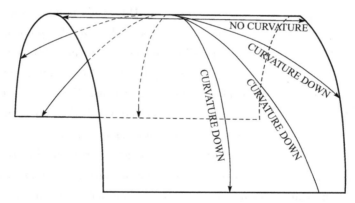

Figure 22-3 Vertical cuts in top half of cylinder.
(Saralinda Hooker and Christopher Ragus)

Words and phrases

1. category   *n.*   类别
2. domelike   *adj.*   穹顶状的
3. cylinderlike *adj.*   柱状的
4. saddlelike   *adj.*   马鞍状的
5. spine   *n.*   脊柱
6. yardstick   *n.*   码尺
7. symmetrical   *adj.*   对称的
8. gutter   *n.*   檐沟,排水沟

Section 2
Extensive Reading

Félix Candela

Félix Candela (January 27, 1910 ~ December 7, 1997) was a Mexican architect and engineer. He worked from the 1930s to the 1960s, and he accomplished a great deal for Mexican architecture (Figure 22-4).

Figure 22-4

Candela's major contribution to structural engineering was the development of thin shells made out of reinforced concrete. He worked very hard during his life time to prove the real nature and potential reinforced concrete had in structural engineering. Reinforced concrete is extremely efficient in a dome or shell like shape. This shape minimizes the tensile forces in the concrete. He also looked to solve problems by the simplest means possible. In regards to shell design, he tended to rely on the geometric properties of the shell for analysis, instead of complex mathematical means. Candela was married to Eladia Martin when he moved to Mexico from Spain; there, they raised a family. Candela was interned as a POW in the Perpignan concentration camp during the Spanish Civil War of the 1930s. In his early life, Felix was active in sports, particularly rugby and skiing. In his later years, his distinguishing feature was his beard that made him look years younger than his true age (since it did not grey a lot). Candela has said on more than one occasion that the analysis of a structure is a sort of "hobby" to him.

As a young adult, Félix Candela was interested in many different fields of study. Due to some advice from a good friend he decided to study architecture. Candela attended the Madrid School of Architecture. Early after he started classes, he developed a very keen sense of geometry and started teaching other students in private lessons. In his junior year, his visual intelligence and his descriptive geometric and trigonometric talent helped him catch the eye of Luis Vegas. Vegas was his material strength professor, and gave Candela the honorary title of "Luis Vegas' Helper". While "helping" Vegas, Candela entered many architecture competitions and won most of them. Unlike many of his peers, Candela didn't show intellectual or aesthetic efforts in school. He didn't even like mathematics. When Candela was a student in Madrid, the schools taught the theory of elasticity. This was a huge problem area for architects, but it didn't phase Candela, who assisted the professors and even tutored other students.

Félix Candela traveled the world, winning many distinguished awards in his career. The reasons for his initial travels from Spain to Mexico were not in pursuit of his career, though in his career this played a large part of his success as an architect/engineer/contractor. Felix Candela was born in Madrid, Spain in 1910. In 1927 Candela enrolled in La Esquela Superior de Arquitrectura, graduating in 1935; at which time Candela traveled to Germany to further study architecture. His studies ended very quickly when the Spanish civil war began in 1936. When Candela returned to Spain to fight, he sided with the republic and fought against Franco. Candela became a Captain of Engineers for the Spanish republic after a short period of time. Unfortunately, while participating in the civil war, Candela was imprisoned in the Perpignan Concentration camp in Perpignan, France until the end of the war in 1939. Candela had fought against Franco; therefore he could not stay in the new Spain as long as Franco was the head of state. Candela was put onto a ship bound for Mexico, where he would start his career.

Resources for Reference

http://en.wikipedia.org
http://www.structurae.de
http://www.eb.com

Section 3
Tips for Translation

Translating meaning from a different point of view 表达方式和视角的转换

在英译汉的过程中,如果遇到某些对译或硬译有困难的情况,不妨改变一下思路,通过表达方式或者视角的转换,选择恰当的译法。

1. The works of some contemporary architects are *beyond my comprehension*.
我看不懂某些当代建筑师的作品。
把"beyond my comprehension"硬译为"超出了我的理解"并不妥当,不符合汉语的表达习惯。

2. As we shall see, *no* large roof can be built by means of natural or man-made compressive materials *without* giving the roof a curved shape, and this is why domes were used before any other type of cover to achieve large enclosed spaces.
就如我们所知,通过普通方式或人工抗压材料建造大型屋顶是不可能的,必须采用曲面,这也是穹顶最早被用来覆盖大型围合空间的原因。

　　本句中的"no"、"without"如果都翻译成"没有"的话，汉语会显得比较怪异，不如采用所谓的"反正表达法"，即把否定形式的词、短语或句子翻译成肯定形式，反而顺畅。

3. There is perhaps *no better way* to understand a great architecture by experiencing it.
理解伟大建筑的最好方式是亲身体验。

　　显然"是……的最好方式"比"没有比……更好的方式"更为自然。

4. Only after the invention of inexpensive methods of steel manufacture and the recent development of reinforced concrete did large flat roofs become possible.
经济的钢材生产方式的发明和钢筋混凝土的最新发展使大型平屋面的建造成为可能。

　　如果在本句的翻译中硬要对应"only after…"翻译成"仅仅当……之后"，反而使翻译缺乏流畅性，不如删繁就简，直接翻译为一般肯定句式。

Unit 23

Section 1
Intensive Reading

Form-Resistant Structures
Mario Salvadori

Part V

BARREL ROOFS AND FOLDED PLATES

We have seen that cylinders are developable surfaces and, as such, are less stiff than either domes or saddles. Even so, they can be used as roofs. Actually barrel roofs of reinforced concrete in the shape of half-cylinders with curvatures down are commonly and inexpensively used in industrial buildings (Figure 23 – 1), since they can be poured on the same cylindrical formwork, which can be moved from one location to another and reused to pour a large number of barrels on the same form.

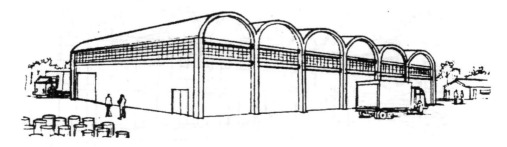

Figure 23 – 1   Barrel roof. (Saralinda Hooker and Christopher Ragus)

The mode of support of a barrel influences its load-carrying action. If a barrel is supported all along its two longitudinal edges [Figure 23 – 2 (a)] it acts as a series of arches built one next to the other and develops out-pushing thrusts, which must be absorbed by buttresses or tie-rods as in any arch. But if it is supported on its curved ends [Figure 23-2 (b)], it behaves like a beam, developing compression above the neutral axis and tension below and it does not develop thrust. One should not be fooled by the geometrical shape of a structure in deciding its load-carrying mechanism. Barrels should be supported on endwalls or stiff arches so as to avoid unnecessary and costly buttresses or interfering tie-rods.

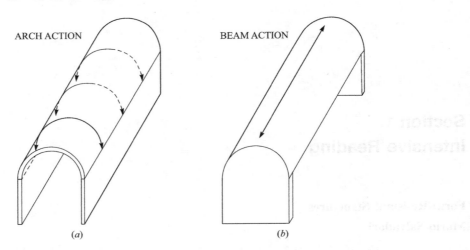

Figure 23-2 Barrel roof supports. (Saralinda Hooker and Christopher Ragus)

The folded plate roof is analogous to a series of barrels. It consists of long, narrow inclined concrete slabs, but presents a sudden change in slope at regular intervals (Figure 23-3). Its cross-section is a zigzag line with "valleys" and "ridges". The construction of a folded plate roof requires practically no formwork, since the flat slabs can be poured on the ground and jointed at the valleys and ridges of the roof by connecting the transverse reinforcing bars of the slabs and using a good cement grout or mortar to make the slabs into a monolithic roof.

Figure 23-3 Folded-plate roof. (Saralinda Hooker and Christopher Ragus)

Folded plates carry loads to the supports along a twofold path. Because of the stiffness achieved by the folds, any load acting on a slab travels first up the nearest ridge or down the nearest valley, and then is carried to the end supports longitudinally by the slabs acting as beams (Figure 23-4). Folded plates must be supported at their ends. Since they consist of flat surfaces and folds, they act like an accordion that can be pushed in or pulled out with little effort, and do not develop outpushing thrusts.

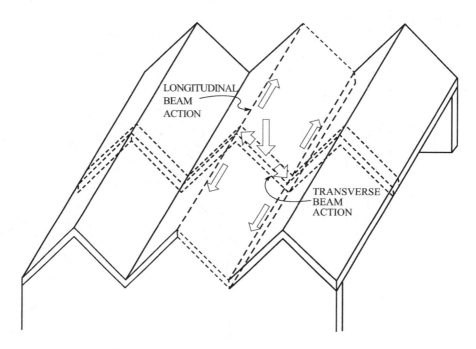

Figure 23-4 Folded-plate load paths. (Saralinda Hooker and Christopher Ragus)

It is both easy and instructive to fold a sheet of thin paper up and down, shaping it into a folded plate, and to support it between two books, possibly laying a flat sheet of paper over it (Figure 23-5). The load capacity obtained by such a flimsy piece of material through its folds is amazing: a sheet of paper weighing less than one-tenth of an ounce may carry a load of books two or three hundred times it own weight! Any reader inclined to experiment further with folded paper can take advantage of both folding and arch action by creasing a sheet of paper into a folded barrel, according to the instructions of Figure 23-6. The creased paper barrel requires buttresses to absorb its outward-acting thrusts, but its load-carrying capacity is even greater than that of a folded-plate roof and may easily reach 400 times its own weight.

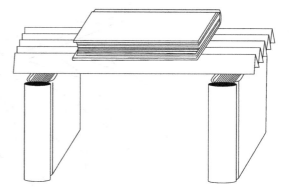

Figure 23-5 Folded-plate paper model. (Saralinda Hooker and Christopher Ragus)

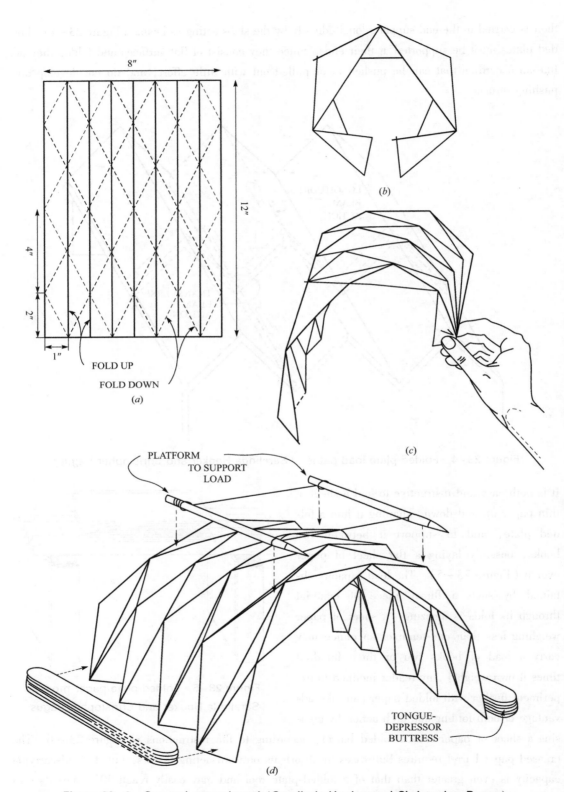

Figure 23-6 Creased paper barrel (Saralinda Hooker and Christopher Ragus)

Words and phrases

1. barrel   *n.*   桶
2. folded plate   折板
3. formwork   *n.*   支模材料
4. longitudinal   *adj.*   纵向的
5. thrust   *n.*   推力
6. tie-rod   *n.*   系杆
7. neutral   *adj.*   中性的，中立的
8. tension   *n.*   张力，拉力
9. endwall   *n.*   端墙
10. analogous   *adj.*   类似的
11. interval   *n.*   间隔
12. transverse   *adj.*   横断的
13. grout   *n.*   水泥浆
14. twofold   *adj.*   双重的
15. flimsy   *adj.*   脆弱的
16. crease   *n./v.*   褶皱
17. visualize   *v.*   形象化
18. inclined   *adj.*   倾斜的

Section 2
Extensive Reading

Passage 1
City Planning According to Artistic Principles (I)
Camillo Sitte
Translated by George R. Collins and Christiane Crasemann Collins (London: Phaidon Press, 1965): 91-104; 105-112

Modern systems! Yes, indeed! To approach everything in a strictly methodical manner and not to waver a hair's breadth from preconceived patterns, until genius has been strangled to death and *joi de vivre* stifled by the system——that is the sign of our time. We have at our disposal three major methods of city planning, and several subsidiary types. The major ones are the *gridiron system*, the *radial system*, and the *triangular system*. The sub-types are mostly hybrids of these three. Artistically speaking, not one of them is of any interest, for in their veins pulses not a single drop of artistic blood. All three are concerned exclusively with the arrangement of *street patterns*, and hence their intention is from the very start a purely technical one. A network of streets always serves only the purposes of communication, never of art, since it can never be comprehended sensorily, can never be grasped as a whole except in a plan of it.... They are of no concern artistically, because they are inapprehensible in their entirety. Only that which a spectator can hold in view, what can be seen, is of artistic importance, for instance, the single street or the individual plaza.

It follows simply from this that under the proper conditions an artistic effect can be achieved with whatever street network be chosen, but the pattern should never be applied with that really brutal ruthlessness which characterizes the cities of the New World and which has, unfortunately and frequently, become the fashion with us.

Artistically contrived streets and plazas might be wrested even from the gridiron system if the traffic expert would just let the artist peer over his shoulder occasionally or would set aside his compass and drawing board now and then. If only the desire were to exist, one could establish a basis for peaceful coexistence between these two. After all, the artist needs for his purpose only a few main streets and plazas; all the rest he is glad to turn over to traffic and to daily material needs. The broad mass of living quarters should be businesslike, and there the city may appear in its workclothes. However, major plazas and thoroughfares should wear their "Sunday best" in order to be a pride and joy to the inhabitants, to awake civic spirit, and forever to nurture great and noble sentiment within our growing youth. This is exactly the way it is in the old towns. The overwhelming majority of their side streets are artistically unimportant, and only the tourist in his exceptionally predisposed mood finds them beautiful, because he likes everything he sees. Just a few thoroughfares and major plazas in the centers of towns stand up under critical appraisal——those upon which our forefathers lavished wisely, and with all means at their disposal, whatever they could muster of works of civic art.

The artistic possibilities of modern systems of city planning should be judged from this standpoint, viz., that of a compromise, since it has already been made quite clear that the modern point of view rejects all demands made in the name of art. Whoever is to be spokesman for this artistic attitude must point out that a policy of unwavering adherence to matters of transportation is erroneous, and furthermore that the demands of art do not necessarily run contrary to the dictates of modern living (traffic, hygiene, etc.)....

The grid plan is the one most frequently applied. It was carried out already very early with an unrelenting thoroughness at Mannheim, whose plan looks exactly like a checkerboard; there exists not a single exception to the arid rule that all streets intersect perpendicularly and that each one runs straight in both directions until it reaches the countryside beyond the town. The rectangular city block prevailed here to such a degree that even street names were considered superfluous, the city blocks being designated merely by letters in one direction and by numbers in the other. Thus the last vestige of ancient tradition was eliminated and nothing remained for the plan of imagination or fantasy. Mannheim assumes the credit for the invention of this system. *Volenti not fit injuria* (No injury is done to a consenting party). One could fill volumes recording the censure and scorn that have been lavished upon its plan in innumerable publications.

In the light of what we have seen it is hard to believe that this very system could have conquered the world. Wherever a new town extension is being planned this method is applied——even in radial and triangular systems the subsidiary nets of streets are organized this way in so far as possible. It is the more remarkable because this very arrangement has long been condemned from the point of view of traffic; Baumeister contains all that has been said to date on the matter. Aside from the inconveniences which he mentions, only one more, which seems to have been overlooked, will be pointed out here, namely the disadvantage of its street crossings for vehicular traffic. To this end let us first examine the traffic pattern where merely one street opens into another (Figure 23-7). In the illustration it is assumed that traffic drives on the left. Thus a vehicle driving from A to C may pass another from C to A or from C to B, as well as from B to A and B to C. These make four encounters. In addition four encounters arise with a vehicle driving from A to B. With vehicles driving from B to A there are only two new encounters. The two others drop out because they are already contained in the earlier series, for it is the same if a vehicle driving from B to A encounters one driving from A to B, or the reverse. Likewise with the vehicles driving from B to C only two new encounters results, and with driving directions C to A and C to B there are no new variants. Avoiding duplications, then, the following twelve situations are possible:

A B   against B A
* A B   against B C
* A B   against C A
A B   against C B
A C   against B A
A C   against B C
A C   against C A

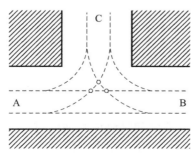

Figure 23-7   Intersecting Streets

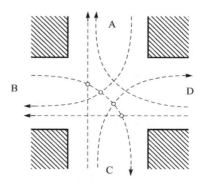

Figure 23-8   Crossing Streets

A C　against C B
B A　against C A
B A　against C B
＊B C　against C A
B C　against C B

Passage 2
Who is Santiago Calatrava?

Santiago Calatrava is a world-renowned architect, engineer and sculptor. Already well-known in Europe for his unique design aesthetic, Calatrava is beginning to make a name for himself in the United States. Starting with the Milwaukee Art Museum, he has designed a number of public buildings and bridges in the U. S. in recent years. As both Engineer and Architect, his works take materials like concrete, glass and steel beyond the normal bounds.

Like many of his buildings, Calatrava's body of work is in constant motion. The images and news items below are updated throughout the day so that you can keep up with the latest Calatrava news. Subscribe to the RSS feed with your favorite RSS reader to be kept up to date at all times.

The PATH Terminal at the WTC site in New York is the final building design for the rebuilding project. In the days immediately following the reveal, it is clear that this building has captured the favor of the general public (Figure 23-9).

Figure 23-9

Resources for Reference

http://www.library.cornell.edu/Reps/DOCS/sitte.htm

http://www.calatrava.info

Section 3
Tips for Translation

Translating meaning 意译

　　翻译的主旨之一是"达意"。如果过于拘泥于英语和汉语词和词的对译，容易流于死板，甚至不知所云。特别是英语中不少约定俗成的俗语，巧妙的翻译往往能起到言简意赅、一针见血的效果。

1. The young are sometimes criticized by the older generation for the "*live now, pay latter*" life style.

年轻人的生活方式有的时候被老一代批评为"寅吃卯粮"。

　　"live now, pay latter" 如果直译，是很难找到恰如其分的文字的。"寅吃卯粮"是汉语成语，生动地传达了英语原文的含义。

2. The non-art-loving public at large, instead of being grateful to architects for what they do, regards the onset of modern buildings and modern cities everywhere as an inevitable, rather sad piece of the larger fact that the world is *going to the dogs*.

大多数不谙艺术的老百姓并不感谢建筑师的所作所为。他们把到处出现的现代建筑和现代城市视为我们这个世界正在不可避免地走向衰落的糟糕现实的一部分。

　　"going to the dogs" 是"堕落"的意思，直译是无法翻译的。

3. ...there is no *silver lining without a cloud* and the cloud that hangs over the hypars is the cost of their framework, as it is for all curved surfaces.

……光明和黑暗总是如影随形。而笼罩在hypars（双曲抛物面）的"黑暗"，就是结构框架的造价。对所有的曲面结构来说都有这样的问题。

　　"silver lining with a cloud" 在英语中是"困境中的一线希望"或"黑暗中的一线光明"的意思。例句的翻译也围绕这一含义，结合汉语的表达习惯进行翻译。

4. There is *growing realization* that more attention should be paid to issues of sustainability in urban planning.

越来越多的人认识到，在城市规划中应当更多地关注可持续发展的问题。

　　"growing realization" 在汉语中很难找到对应的表达，不如意译成"越来越多的人"。

5. We cannot get an *adequate picture* of what Middlesborough is, or of what it ought to be, in terms of 29 large and conveniently integral chunks called neighborhoods.

从29个被称为邻里的生活便利的完整区块我们可能无法完全了解米德尔斯堡究竟如何，或应当如何。

　　"an adequate picture" 可意译为"完全了解"。

6. Take the separation of pedestrians from moving vehicles, is a tree concept proposed by Le Corbusier, Louis Kahn, and many others. At a very crude level of thought this is obviously a good idea.

勒·柯布西耶、路易·康等很多人提出了行人和机动车分离的"树状"规划理念。粗粗一想这显然是个好主意。

　　"at a very crude level of thought" 可意译为"粗粗一想"。

7. Another favorite concept of the CIAM theorists and others is the separation of recreation from

everything else. This has crystallized our real cities in the form of playgrounds.
CIAM 的理论家最喜欢的一个概念是把休憩功能同其他功能分离。游戏场是这一概念在城市中的具体体现。

"in the form of" 本意是 "以……为形式"，在此结合 "crystallize" （使……具体化）的意思，翻译成 "是……的具体体现"。

8. In Cambridge, a natural city where university and city have grown together gradually, the physical units overlap because they are the physical residues of city systems and university systems which overlap.
剑桥是一个大学和城市一步步共同成长的自然城市。物质要素相互交叠，因为这些要素是城市和大学相互交叠后的产物。

"residue" 本意是 "（化学反应后）残留物" 的意思，在此可意译为 "产物"。

Unit 24

Section 1
Intensive Reading

Form-Resistant Structures
Mario Salvadori

Part VI

SADDLE ROOFS

Saddle surfaces, supported along their longitudinal curved edges, have a particularly elegant shape which blurs the distinction between structure and functional skin (Figure 24–1). But saddle surfaces make some of the loveliest roofs when cut and supported along those straight lines which we have seen necessarily exist on any surface with both up and down curvatures. To visualize how a curved surface can be obtained by means of straight lines, connect by inclined straight-line segments the points of two equal circles set one above the other (Figure 24–2). The segments generate a curved surface called a rotational hyperboloid, used to build the enormous cooling towers of chemical plants. One of the most commonly used roof surfaces is obtained in a similar manner. Imagine a

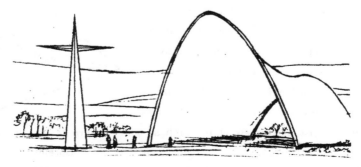

Figure 24–1 Candela's saddle-shaped chapel in Mexico.
(Saralinda Hooker and Christopher Ragus)

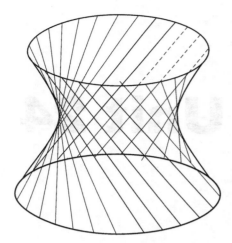

Figure 24-2 Rotational hyperboloid. (Saralinda Hooker and Christopher Ragus)

rectangle of solid struts, in which one of the corners is lifted from the plane of the other three, thus creating a frame with two horizontal and two inclined sides [Figure 24-3 (a)]. If the corresponding points of two opposite sides of this frame (one horizontal and one inclined) are connected by straight lines, for example by pulled threads, and the same is done with the other two opposite sides, the threads will describe a curved surface, although, being tensed, they are themselves straight [Figure 24-3 (b)]. This surface has a curvature up along the line connecting the lifted corner to its diagonally opposite corner, and a curvature down the direction of the line connecting the other two corners (Figure 24-4). It is, therefore a saddle surface. It carries the high-sounding name of hyperbolic paraboloid, wisely shortened to hypar by our British colleagues.

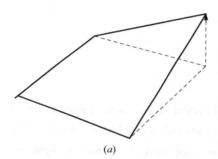
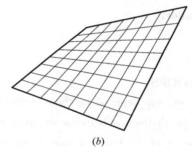

Figure 24-3 Hypar frame (a) and hypar straight-lines surface (b). (Saralinda Hooker and Christopher Ragus)

One of the simplest hypar roofs is obtained by tilting the saddle and supporting it on two opposite corners. Whether the support points are on the ground or on columns, the roof looks like a butterfly ready to take off (Figure 24-5). Its structural behavior is dictated by its curvatures. Compressive arch action takes place along the sections curved downward and tensile cable action along the sections curved upward. The two support points must be buttressed to resist the thrusts of the arch action, while the tensile cable action at right angles to it must be absorbed by reinforcing bars, if the hypar is made out of concrete. Such is the stiffness of a hypar that its thickness need be only a few inches

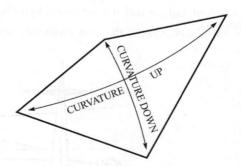

Figure 24-4 Hypar curvatures. (Saralinda Hooker and Christopher Ragus)

of concrete for spans of thirty or forty feet. The hypar has other wonderful structural properties. For example, one could fear that such a thin structure, acting in compression along its arched direction, would easily buckle, a fear quite justified were it not that the cable action at right angles to the arches pulls them up; prevents them from buckling! Finally, to make the structural engineer even more enamored of these surfaces, under a uniform load, like its dead load or a snow load, they develop the same tension and compression everywhere. Therefore, its material, be it concrete or wood, can be used to its greatest allowable capacity all over the roof. The reader, who might not have seen too many of these magnificent roofs, may ask "Why do we not see many more of them?" The answer to this question is that there is no silver lining without a cloud and the cloud that hangs over the hypars is the cost of their framework, as it is for all curved surfaces. More will be said later about this problem.

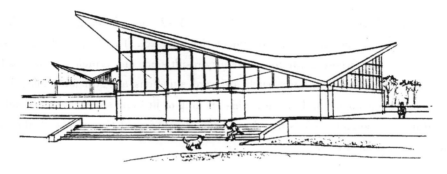

Figure 24-5　Hypar butterfly roof. (Saralinda Hooker and Christopher Ragus)

Words and phrases

1. rotational hyperboloid　旋转双曲面
2. cooling tower　冷却塔
3. strut　*n.*　压杆
4. corresponding　*adj.*　相应的
5. pulled thread　拉索
6. describe　*v.*　描述，画
7. diagonally　*adv.*　对角地
8. high-sounding　*adj.*　夸张的
9. hyperbolic paraboloid　双曲抛物面/hypar（缩写）
10. tilt　*v.*　（使）倾斜
11. buckle　*v.*　弯曲，变形，起皱
12. justify　*v.*　证明……是适当的
13. be enamored of　迷恋于……
14. dead load　静荷载/live load 动荷载
15. magnificent　*adj.*　宏伟的

16. framework *n.* 框架
17. silver lining（不幸或失望中的）一线希望

Section 2
Extensive Reading

Passage 1
City Planning According to Artistic Principles (Ⅱ)
Camillo Sitte
Translated by George R. Collins and Christiane Crasemann Collins (London: Phaidon Press, 1965): 91-104; 105-112

By checking each of these twelve encounters in Figure. 23-7, one can easily see that those designated with an asterisk are such that their two trajectories intersect. The little circles represent these three bad traffic situations wherein a traffic delay may result since one vehicle has to pass by before the other can proceed. However, three such awkward situations are still allowable, because in light traffic a congestion would rarely result. This opening of just one street into a second (usually a broader and more important one) is the most common case in old towns, and at the same time the most advantageous for traffic.

The situation is much worse when streets actually cross completely (Figure. 23-8). Here the various encounters, diagrammed and calculated without duplication, come to 54, among which occur 16 cases of intersecting traffic trajectories. This is more than five times as many crossings and possible traffic disruptions as before. The course of a single vehicle driving from B to C is cut by four others, and the vehicle moving from C to D comes at it exactly from the side. Therefore at such crossings, when they are very busy, drivers must go at a slow pace, and anyone who drives about much in carriages knows that in the modern sections of town he is often slowed down, while in the narrow alleys of the old part of town, crowded with traffic as they are, he can proceed quite nicely at a trot. This is reasonable because a street seldom crosses there, and even simple street openings are relatively infrequent.

For pedestrians the situation is even worse. Every hundred steps they have to leave the sidewalk in order to cross another street, and they cannot be careful enough in looking to the right and left for vehicles which may be coming along every which way. They miss the natural protection of uninterrupted house fronts. In every town where a so-called corso or promenade has developed, one can observe how a long continuous row of houses was instinctively chosen as side-protection, since otherwise its whole pleasure of strolling would be spoilt by the constant lookout for cross traffic....

But what marvelous traffic conditions arise when more than four thoroughfares run into each other! With the addition of just one more street opening to such a junction, the possible vehicle encounters already total 160 which is more than ten times the first case, and the number of crossings which disrupt traffic increases proportionately. Yet what shall we say about traffic intersections where as many as six or more streets run together from all sides, as in Figure 24-6? In the center of a populous town, at certain busy times of day, a smooth flow of traffic is actually impossible, and the authorities have to intervene, first, by stationing a policeman who, with his signals, keeps the traffic precariously moving. For pedestrians such a place (Figure 24-7) is truly hazardous, and in order to eliminate the worst dangers, a round piece of sidewalk is raised in the middle——a small safety island on which a beautiful slender gas light rises like a lighthouse amidst the stormy waves of the ocean of vehicles. This safety island with its gas lamp is perhaps the most magnificent and original invention of modern city planning! In spite of all these precautions, crossing the street is advisable only for alert persons; the old and the frail will always by preference take a long detour in order to avoid it.

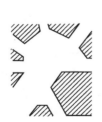

Figure 24-6 Cassel:
Kölnerstrasse

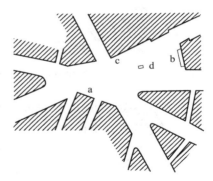

Figure 24-7 London: a. Mansion House b. Stock Exchange. c. Bank of England. d. Wellington Statue

These, then, are the achievements of a system that, relentlessly condemning all artistic traditions, has restricted itself exclusively to questions of traffic. Its monstrous street junctions are called "plazas", yet in them everything is avoided that would make for character in a plaza, and at the same time everything seems to be accumulated that is impractical and ugly. These are the consequences of design based on traffic considerations rather than, as it should be, on the arrangement of plazas and streets. In the gridiron system junctions like this result wherever the difficulties of terrain or a need to relate to what already existed before require deviations or breaks in the checkerboard pattern; triangular so-called plazas come into existence, as seen in Figure 24-8 ~ Figure 24-10. These occur even more frequently in the application of the radial system or in mixed systems. (Figure 24-11) They become the greatest glory indeed of new layouts when they are completely regular: in circular form (Figure 24-12), or octagonal as in the Piazza Emmanuele in Turin. Nowhere can the bankruptcy of all artistic feeling and tradition be more clearly perceived than

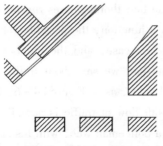

Figure 24-8 Trieste: Piazza della Caserma (Guglielmo Oberdan)

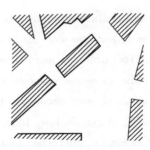

Figure 24-9 Trieste: Piazza della Legna (Carlo Goldoni)

Figure 24-10 Trieste: Piazza della Borsa

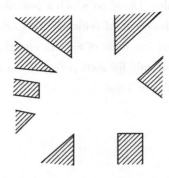

Figure 24-11 Lyons: Place du Pont

Figure 24-12 Cassel: Königsplatz

here. In plan such a plaza appears, of course, to be nicely regular, but what is the consequence in reality? Vistas opening out along a thoroughfare, which the ancients avoided so artfully, have here been used as much as possible The traffic junction is also a junction of all lines of sight. As one circles the plaza he always sees the same panorama, so that it is never exactly clear where one is standing A stranger has only to turn around once on such a disconcerting merry-go-round of a plaza and immediately all sense of orientation is lost. On the Piazza Vigliena (Quattro Canti) in Palermo even the elaborate decoration of the four corners does not help, since they are all alike. Although only two major streets intersect perpendicularly on this octagonal plaza, one still finds strangers frequently turning into one of them to look for the street name or a familiar house, thus to regain their orientation. In reality all that is attained is a complete loss of our bearings, a monotony of vistas, and an architectural ineffectiveness. How odd a whim of the old masters to have ascribed importance to the avoidance of such things!

Passage 2
Public domain

Public space between buildings influences both the built form and the civic quality of the city, by streets, squares or parks. A balance between the public and private domain is central to the practice's design approach. Buildings and their surrounding spaces should interrelate and define one another, with external spaces functioning as rooms without roofs.

It is the celebration of public space, and the encouragement of public activities that drives the form of the practice's buildings. It is the building's scale and relationship with the street or square that helps to encourage public activity and create a people-friendly environment. For example, the steps that lead to the Channel 4 Headquarters, the narrow passage that runs around the Lloyd's of London building, the small churchyard in front of Lloyd's Register, the close around the National Assembly for Wales or the square in front of the Bordeaux Law Courts are all examples where the relationship between buildings and public spaces demonstrate how the architect's responsibility can successfully extend beyond the brief to include the public domain.

The Pompidou Centre in Paris, designed by Richard Rogers in collaboration with Renzo Piano and completed in 1977, illustrates how a building can bring life to a rundown area of a city. The design deliberately dedicated over half of the site to a public piazza. The public domain, in this case, extends from the square up the facade of the building in the form of "a street in the air", a great diagonal escalator crossing the facade to connect all the floors. The Pompidou Centre, including its piazza has become the most visited building in Europe with spontaneous street theatre and other events in the piazzas complimenting the activities within the building.

The concern for public space continues to be firmly embedded in the vision for the practice's work. In a recent project located in the City of London, for a high-rise office building at 122 Leadenhall Street, seven of its 48 storeys are dedicated to cafés, restaurants and other public facilities within a large south-facing volume connected to an existing plaza which flows through the building uninterrupted at ground level.

The enormous scale of this space is unprecedented in London, and will become the focal point for the project as well as a major new meeting space contributing to the vibrant life of the City. The project is a fine example of how private and public domain can combine to contribute to the quality of our cities (Figure 24–13、Figure 24–14).

Figure 24–13

Figure 24-14

Resources for Reference

http://www. library. cornell. edu/Reps/DOCS/sitte. htm
http://www. richardrogers. co. uk/theory/public_domain

Section 3
Tips for Oral Presentation

Use Your Everyday Speaking Situations

Whenever you speak to people, make an extra effort to notice how you speak. Observe, too, whether the facial expressions of your listeners indicate they do or do not understand what you are saying. Before calling to request something on the phone, plan and practice what you are going to say. Even this is essentially a short presentation. Another exercise is to prepare a 90-second presentation about yourself. Describe who you are and what you do. Record your presentation and review it using the four steps described above.

Since you are talking about yourself, you don't need to research the topic; however, you do need to prepare what you are going to say and how you are going to say it. Plan everything including your gestures and walking patterns.

Resource for Reference

http://www. ljlseminars. com/bodyspeaks. htm

Unit 25

Section 1
Intensive Reading

Form-Resistant Structures
Mario Salvadori

Part VII

COMPLEX ROOFS

The barrel and the rectangular hypar elements are the building blocks for some of the most exciting curved roofs conceived by man. Combinations of these structurally efficient components are limited only by the imagination of the architect, guided by good structural sense. It is indeed regrettable that, with a few notable exceptions, modern architecture has not used curved surfaces as glorious as some of the past and at the same time as daring as present day technology can make them. This lack of achievement is due to at least three causes. On one hand, curved surfaces are believed to be more complex to design than the flat rectangular shapes we are so used to. Usually quite the opposite is true. On the other hand, there is a gap between recent curved-structures theory and the prescriptions of the codes. A domed roof proposed for a bank in California——meant to cover a rectangular area ninety feet by sixty feet and to be only a few inches thick——was vetoed by the local building department engineer because thin curved roofs were not mentioned in the code and, hence, "did not exist".

The engineer would only allow the roof to be built if two concrete arches were erected between its diagonally opposite corners "to support it". Little did he know that the thin concrete roof was so stiff that it would support the two heavy arches rather than be supported by them. Finally, one must honestly add that in the United States the ratio of labor to material costs often makes *thin shells* (as these curved roofs are usually labeled) uncompetitive with other types of construction. The situation is reversed in Europe and other parts of the world.

One of the most commonly encountered combinations of cylindrical surfaces is the *groined vault* of the Gothic cathedrals (Figure 25-1). This consists of the intersection of two cylindrical vaults at right angles to each other, supported on four boundary arches and intersecting along curved diagonal folds called groins, which end at the four corner columns supporting the vault. The groins have often been emphasized visually and, possibly, structurally by means of ribs but, though these ribs may be aesthetically important, they are not needed to sustain the vaults. By their curvature and folds they are self-supporting.

Among the great variety of combinations of rectangular hypar elements, two have become quite common because of their usefulness, beauty, and economy: the hypar roof and the hypar umbrella. To put together a hypar roof (Figure 25-2), consider building four hypar rectangular elements, starting as was done before with four rectangles but lowering (rather than lifting) one corner in each of them. The hypar roof is obtained by joining together the horizontal sides of each rectangular element so that all eight meet at the center of the area to be covered, while the lowered corners are supported on four columns or on the ground at the corners of the area. The straight inclined sides of the roof act as the compressed struts of a truss, and they must be prevented from spreading outward by means of tie-rods connecting its corners, all around the covered area. The largest roof of this kind has been erected in Denver, Colorado, and measures 112 feet by 132 feet, with a three-inch thickness. It rests directly on the ground at the corners and covers a large department store.

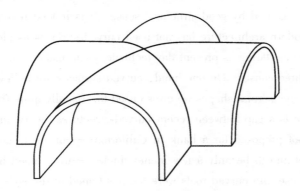

Figure 25-1 Groined vaults.
(Saralinda Hooker and Christopher Ragus)

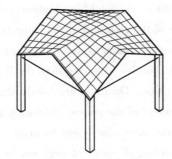

Figure 25-2 Hypar roof:
and hypar umbrella

Words and phrases

1. rectangular *adj.* 矩形的
2. conceive *v.* 构想，设想

3. combination *n.* 组合，结合
4. regrettable *adj.* 可惜的
5. glorious *adj.* 壮丽的，光辉灿烂的
6. prescription *n.* 规定
7. code *n.* 法规，准则
8. veto *v.* 投票反对/vote 投票赞成
9. thin shell 薄壳
10. groined vault 交叉拱
11. Gothic cathedral 哥特式天主教堂
12. intersection *n.* 交叉
13. aesthetically *adv.* 审美地，美学上的

Section 2
Extensive Reading

Passage 1
City Planning According to Artistic Principles (III)
Camillo Sitte
Translated by George R. Collins and Christiane Crasemann Collins (London: Phaidon Press, 1965): 91-104; 105-112

This type of plaza, along with its safety island and gas light or columnar monument, found its earliest manifestation in Paris (Figure 25 – 3) although none of the modern systems we have described happened to be rigorously carried out there during the last big renovation of the city. This was due in part to the intractable nature of the existing layout and in part to the tenacity with which fine old artistic traditions had preserved themselves. Different procedures were followed in various parts of the city, and, if nothing else, one can suggest that a certain remnant of Baroque tradition served as a common basis. The striving for perspective effects has obviously continued, and we could designate as the backbone of the system the broad avenue closed off in the distance by a monumental structure.... Later the modern motif of the ring-boulevard was added to this, and a certain vigorous clearing out or breaking through of the dense mass of old houses was required by the circumstances. This remarkable reorganization, carried out on a large scale, became almost a fad, first and most frequently observable in the large cities of France.

The Place Juan Juares at Marseilles (Figure 25 – 4) should be mentioned as an example of the ruthless carving of a plaza out of a web of crooked streets. The Place du Pont at Lyons (Fig. 86) and other similar ones should also be noted. This practice has something vaguely in common with

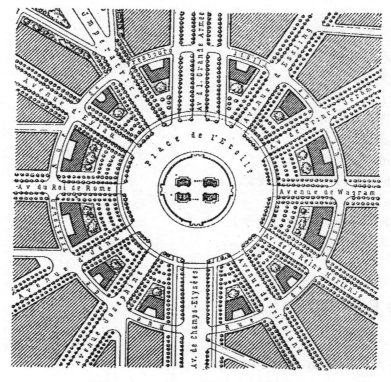

Figure 25-3　Paris: Place de l' Etoile

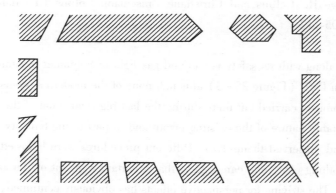

Figure 25-4　Marseilles: Place Juan Juarès (formerly Place St. Michel)

Nero's radical rearrangement of Rome, although, of course, considerably more modern in character. Avenues and ring-boulevards were developed at Marseilles; at Nimes (Cours Neuf, Boulevard du Grand Cours, Boulevard du Petit Cours); at Lyons (Cours Napoleon); at Avignon (Cours Bonaparte); and in other cities. In Italy a similar broad artery with several traffic lanes and shaded walks is called a corso or largo. Broad circumvallating boulevards were usually developed on the circuit of abandoned fortifications——in Vienna, Hamburg, Munich, Leipzig, Breslau, Bremen, Hanover; at Prague between the Altstadt and the Neustadt; at Antwerp; as a pentagon at Wurzburg (Juliuspromenade, Hofpromenade, etc.) and elsewhere. The avenue as a very old and independently developed motif is, for instance, to be found in the Langgasse at Danzig; the Breite Gasse at Weimar; the Kaiserstrasse at Freiburg; the Maximilianstrasse at Augsburg; Unter den Linden in Berlin. The Jagerzeile in Vienna (now Praterstrasse) is representative of such broader avenues developed for their long vistas, and the Graben there will, after its redesigning is completed, be transformed from a plaza into such an avenue. These are forms in modern city planning that are still artistically effective and are truly in the spirit of the Baroque.

However, as soon as the geometric pattern and the building block became dominant, art was forced into silence. The modernizing of Gotha, Darmstadt, Dusseldorf, the fan-shaped plan of Karlsruhe, etc., are examples of this. The absence of pedestrians on so many modern gigantic streets and plazas (the Ludwigstrasse in Munich, the Rathausplatz in Vienna) in contrast to the crowds in the narrow alleyways of the older parts of towns, demonstrates unequivocally how little the matter of traffic received its due consideration in such city expansions, although supposedly everything was based on just that. Whereas new broad streets are laid out on the periphery of the city where dense traffic is never likely to develop, the old city center remains forever congested. This should be proof enough that the exponents of an exclusively traffic-oriented point of view, despite occasional success, are not justified in throwing to the winds as useless the assistance of art, the teachings of history, and the great traditions of city building. One more important motif of modern planning remains to be mentioned. This is the matter of tree-lined avenues and gardens. Without doubt they constitute an important hygienic factor, and they also afford the undeniable charm of landscape elements in the middle of a big city and, occasionally, a splendid contrast between groups of trees and architecture. Yet it is open to question whether they are placed at the right spots. From the purely hygienic aspect the answer seems quite simple: the more greenery, the better——that is it in a nutshell. Not so from the artistic point of view, for the question arises as to where and how the greenery is to be applied. The usual and most felicitous application is to be found in the residential sections of modern cities, as in the justly famous residential belt around Frankfurt a. M., the Cottage Anlagen of the Wahring district in Vienna, the similar annexes to the old part of Dresden, etc., as well as the indispensable villa areas at every spa: Wiesbaden, Nice, etc.

However, the closer such landscape elements encroach upon the center of a large city, and especially upon large monumental structures, the more difficult it becomes to find a universally satisfactory and artistically faultless solution. Modern naturalistic landscape painting is not suitable for monumental purposes; when it is used as a background for great mythological and religious representations or in the interiors of monumental buildings or churches, there necessarily arises an uncomfortable conflict between the Realism of style and the Idealism of subject matter, which no device, however clever, can relieve. In just the same way, the penetration of the English park into the major plazas of a city produces a conflict between the principles and effects of naturalism and the rigor of a monumental style. An awareness of this contradiction and a wish to avoid it were the forces which brought into being the Baroque park with its trimmed trees; an architectonically disciplined nature was used in former times primarily in connection with the chateau, whereas the larger monumental city-squares of classic times, of the Middle Ages, and of the Renaissance were exclusively focal points of creative art and especially of architecture and sculpture. Just how annoying the planting of trees in front of such works can be——above all on shabby, sickly boulevards——is to be seen in any photograph.... Photographs are always of winter views, so that important architecture is at least partly visible between the bare branches; in fact, a drawing is frequently preferred to a photograph because with the former any disturbing trees can be left out entirely. Should they not, for the same reason, better be left out in reality, too? What value does an open plaza have as a perspective space when it is congested with foliage?

Passage 2
Flexibility

Today's buildings are more like evolving landscapes than classical temples in which nothing can be added and nothing can be removed. Open ended, adaptable frameworks with large, well-serviced and well-lit floors, on the other hand, offer the possibility for a long life span for the building and a variety of possible uses. For example, Mossbourne Community Academy and Minami School will be able to adapt over time to progressive approaches to education. This concept was developed in earlier buildings such as Lloyd's of London and the Pompidou Centre, solutions that include spaces that can be used for multiple activities in the short term, as well as having many alternative long term uses depending on future requirements.

At Barajas Airport in Madrid the objective was to ensure that the architecture was robust enough to handle the amount of passengers estimated for forthcoming years. The one constant factor with airports is change, and so the essence of this scheme in constructional terms was its use of a single modular segment for the entire 1.2km long structure, allowing ease of expansion whilst disciplined in response to environmental constants such as day-lighting. Repetition of elements as an aid to the construction process can be seen at Chiswick Park, facilitated by design consistency across the individual buildings.

Office occupiers require flexible spaces in order to respond to contingencies in business life; they need to be able to extend and adapt buildings. A concept that incorporates a high level of standardised design will facilitate change.

For functional reasons we always create clear zoning between servant and served spaces within a building. We often separate and juxtapose the services with the mass of the building; in practical terms the part of the building which is inhabited has a long life, whereas the technical services have a short life and therefore need to be accessible for change and maintenance. By separating the mechanical services, lifts, electrics, fluids and air-conditioning from the rest of the building, inevitable technical developments can be incorporated where they are most needed to extend the life of usable core space. The articulation of the services and core building creates a clear three-dimensional language, a dialogue between served and servant spaces and a means of creating flexible floor space. Standardised large floor-plates with services placed on the perimeter have been successful in commercial buildings such as 88 Wood Street and Lloyd's Register, and allow for flexible tenancies that respond to the changing demands of the office market.

Our masterplans are defined by an holistic approach allowing for enough flexibility to accommodate changes over the lifespan of a city, in order to meet market demands. The practicalities and concept of a public place should be inherently flexible in order to respond to changes such as transport and density.

Our masterplans are predominantly not derived from form alone but are concerned with facilitating the movement of people between places, creating a critical mass for successful people spaces and the potential for land to accommodate different uses over time.

Resources for Reference

http://www.library.cornell.edu/Reps/DOCS/sitte.htm
http://www.richardrogers.co.uk/theory/flexibility

Section 3
Tips for Translation

人名和地名的英译

汉语人名的翻译采用汉语普通话拼音，姓在前，名在后，双名应连在一起，姓和名的第一个字母大写。如邓小平——Deng Xiaoping，诸葛亮——Zhuge Liang。某些历史人物的翻译应采用约定俗成的译法，如孔子——Confucius，孟子——Mencius，孙中山——Sun Yat-sen，等等。

地名翻译也采用汉语普通话拼音，如上海——Shanghai，杭州——Hangzhou，等等。当地名的第一个字拼音以元音结尾，第二个字以元音开头的话，应以"'"分开，避免混淆，如西安——Xi'an。

路名的翻译亦采用汉语普通话拼音，但应当注意的是，在汉语中经常有"……南路"、"……北路"和"南……路"、"北……路"的路名，在翻译此类路名的时候，一般将表示方位的"南"、"北"等翻译成"south"、"north"等，其他情况翻译成"nan"、"bei"等。如苏州南路——South Suzhou Road，南苏州路——Nansuzhou Road。

在翻译地址的时候，一般按照英语的习惯，从小到大，邮政编码放在城市名后。如中国上海四平路1239号同济大学，邮政编码200092，译作Tongji University, 1239 Siping Road, Shanghai 200092, P. R. China。在翻译"……新村"等中国特有名称的时候，一般用汉语拼音，反而清楚明了，避免混淆。如曲阳新村——Quyang Xincun。

Unit 26

Section 1
Intensive Reading

Form-Resistant Structures
Mario Salvadori

Part VIII

The hypar umbrella [Figure 26–1], one of the most elegant roof structures ever devised, is produced by using four rectangular hypar elements, each with a corner lowered with respect to the other three, and put together by joining the two inclined sides of each rectangular element so that all eight meet at the center of the area to be covered. The horizontal sides constitute the rectangular edge of the roof and hide the tie-rods. The shape of this hypar roof, which starts at a central point and opens up, resembles that of a rectangular umbrella and gives a visual impression of floating upward. Hypar umbrellas up to ninety-feet square have been used, for example, in the terminal building of Newark Airport [Figure 26–2].

Such is the variety of shapes which can be composed by means of hypars that the Spanish architect Felix Candela has become famous all over the world by designing and building, mostly in Mexico, roofs that use only this surface as basic element. Even though the hypar has particularly efficient structural properties when used as a horizontal roof, Candela has shown, in Mexico City, how exciting its form can be when be used vertically as in the Iglesia de la Virgen Milagrosa. Nervi also has used vertical hypars as walls and roof in the monumental Cathedral of San Francisco (Figure 26–3).

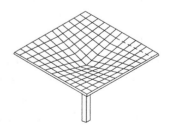

Figure 26–1 Saralinda Hooker and Christopher Ragus

Figure 26-2 Hypar umbrellas of Newark Airport.
(Saralinda Hooker and Christopher Ragus)

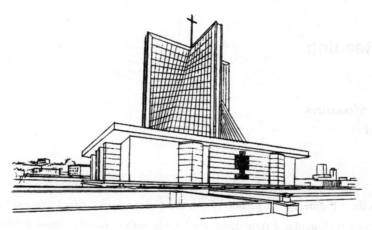

Figure 26-3 Cathedral of San Francisco.
(Saralinda Hooker and Christopher Ragus)

THIN SHELL DAMS

The greatest application of vertical, concrete thin shells has come not in architecture, however, but in dam construction. Although the world as a whole has so far only utilized 15 percent of the power obtainable from its natural or artificial waterfalls, the [former] USSR has exploited 18 percent of their potential, the United States 70 percent and the three European Alpine countries (France, Switzerland, and Italy) 90 percent. Dams can be built to contain water by erecting a heavy wall of earth at the end of a valley and compacting it so as to make it watertight (Figure 26-4). These dams resist the horizontal pressure of the water behind them by means of their weight (as the weight of a building resists the horizontal pressure of the wind) and are called *gravity dams*. They are commonly used in developing countries, where labor is abundant and inexpensive and heavy earthmoving equipment rarely available. On the other hand, where valleys are deep and their sides are formed by rocky mountains, and where concrete technology is well developed, dams are often built as thin, concrete, curved surfaces, which resist the pressure of the water through their curvature (Figure 26-5). (They may be thought of as curved roofs loaded with snow, but rotated into a vertical position, so that the snow load becomes horizontal.) Some of the Alpine dams are monumental structures reaching heights of over 1.000 feet and transmitting the thousands of tons of water pressure to the valley sides through their curvature. It may be thought ironic that such structures be called "thin", when their thickness, which increases from top to bottom, may reach ten feet. But thickness is never measured in absolute terms: what counts structurally is the ratio of the thickness to the radius of the curved surface, which in a dam can be as low as 1/500. It can be realized how "thin" a ten-foot-thick dam is by comparing it with the curved shell of an egg, in which the thickness to radius ratio is as much as 1/50. A dam is, relatively speaking, ten times thinner than an eggshell.

Figure 26-4 Gravity dam.
(Saralinda Hooker and Christopher Ragus)

Figure 26-5 Thin-shell dam.
(Saralinda Hooker and Christopher Ragus)

One of the questions often asked of the structural engineer is whether any of the beautiful curved surfaces encountered in nature or imagined by the fertile mind of an artist could be used to build roofs or other structures. For example, one lovely, thin-shell roof in California has been built in the shape of a square-rigger sail, blown out by the wind but then turned into a horizontal position and supported on its four corners (Figure 26-6). Although not a geometrically definable surface, it is structurally efficient. On the other hand, though an undulating surface can have a pleasant appearance, it would be quite inefficient structurally due to its tendency to fold like an accordion. We can learn a lot from nature, only if we know how to look at it with a wise and critical eye.

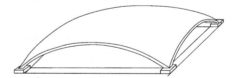

Figure 26-6 Thin-shell "sail" roof. (Saralinda Hooker and Christopher Ragus)

Words and phrases

1. Alpine country (France, Switzerland, and Italy)　阿尔卑斯山国家（如法国、瑞士、意大利等）
2. watertight　*adj.*　水密的/waterproof　防水的
3. gravity　*n.*　重力

4. ironic *adj.* 讽刺的
5. radius *n.* 半径/diameter 直径
6. fertile *adj.* 肥沃的，丰富的/sterile 贫瘠的
7. square-rigger sail 横帆船
8. undulating *adj.* 起伏的

Section 2
Extensive Reading

Passage 1
City Planning According to Artistic Principles（Ⅳ）
Camillo Sitte
Translated by George R. Collins and Christiane Crasemann Collins（London：Phaidon Press, 1965）：91-104；105-112

From this derives the principle that trees should not be an obstruction to the line of sight, and this rule in itself requires a return to Baroque models.

Complete adherence to this strictly artistic principle is impossible in modern city planning since it would put an end to almost all tree planting. Just as for monuments, we have no proper place for trees. The cause of this evil is the same in both instances——namely the modern building block. It is quite astonishing how many delightful small gardens are to be found in the interior of the building lots of old towns; one has no suspicion of their existence before entering the courtyards and rear areas. What a difference between these small private gardens and most of our public parks today! The old private garden, customarily connected with several adjacent ones, all of which are guarded from the wind and dust of the streets by the enclosing facades of high buildings, provides a really refreshing coolness and, insofar as possible in the big city, clean dust-free air. It is truly a garden for the relaxation of the owner, and it is a blessing for all the surrounding interior apartments which thereby receive better air, daylight, and a pleasant view into the greenery. In contrast to this, the interior room of a modern apartment building——with its view into narrow, stuffy, dark and frequently bad-smelling courts, so filled with stagnant air that the windows are forced to remain shut——is a dungeon of the most disagreeable sort, which repels all tenants and increases the demand for outside apartments, much to the detriment of our building projects. The modern public garden, being surrounded by open streets, is exposed to wind and weather and is coated with street dust, unless somewhat protected by its enormous size. Thus it happens that all these open modern parks fail completely in their hygienic purpose and are actually shunned by the public because of their dirt and heat, especially during the warm summer-time.

The fundamental reason for this is again the abominable block system, since gardens should also, just like buildings and monuments, follow the example of the ancients——not standing free in the middle of empty spaces but being built in. As an example of such an inappropriate planting of trees, the plaza behind the new Bourse in Vienna should be mentioned.

Hygienically speaking it is certainly quite indifferent whether these few trees stand there or not, since they provide neither shade nor coolness; rather they can scarcely be prevented from dying on account of the heat and dust. They only succeed in obstructing the view of the Bourse building. Would it not be better then to save the useless waste of a meager tree planting at such a place and instead create real gardens which for their own preservation are enclosed and, most important, do not lie open to the street? Wherever private gardens that formerly belonged to palaces have been given over to public use, one can see for oneself that, secluded from traffic, such gardens fulfill their hygienic purpose despite their small size and vegetation thrives. The uselessness of widely spaced greenery on streets and especially on public walks is demonstrated clearly enough by the fact that on hot summer days promenaders saunter not on these walks, but along the footways of the Ringstrasse, the avenues, etc. The principal value would consist in spraying the leaves with water so that during heatspells they could serve as an evaporating, and hence an air conditioning apparatus. Even this slight profit would be enough to justify street greenery wherever possible. However, in front of monumental structures the file of trees should be interrupted, since here aesthetic considerations are certainly much more important than their small hygienic value, and, as the lesser of two evils, the row of trees should be broken.

That this schism between the old and the new approach also occurs in the field of horticulture allows us now to recapitulate. The effective enclosure of space, deriving as it does from the historical evolution of an original unbroken street front (such as still exists today in villages), continued to be the basis of all dispositions in the old towns. Modern city planning follows the opposite tendency of dissection into separate blocks——building block, plaza block, garden block——each one being clearly circumscribed by its street frontage. From this develops a powerful force of habit: the desire to see every monument in the center of a vacant space. There is method in this madness. The ideal behind this planning could be defined mathematically as a striving for the maximum of frontage line, and herein would appear to lie the creative impulse behind the modern block system. The value of every building site increases with the length of its street frontage, the maximum value for building lots in the parcelling of land being therefore achieved when the perimeter of each block of buildings is greatest in relation to its area. Thus from a purely geometrical point of view circular blocks of buildings would be the most favorable, and, indeed, in the same configuration as that in which large balls of equal size can be pushed the closest together, namely six around one in the middle. In arranging straight streets of identical width between such blocks, the circular forms would be transformed into regular hexagons, as used in tile patterns or in the honeycomb. One could not believe it humanly possible that an idea of such really oppressive ugliness, of such appalling tediousness, and of such a labyrinthine lack of orientation would actually be carried out. Yet, incredible as it seems, it has become a reality in Chicago.

That then is the essence of the block system! In it art and beauty are no more. To arrive at such extremes is impossible in the Old World where we are used to the beauty and the coziness of old towns. However, many of their charms are already irretrievably lost for us since they no longer harmonize with modern living. If we do not wish to let this situation get out of hand, but still want to save as much as possible of artistic value in the layout of cities, we must be clear in our own minds as to just what can still be retained and what has to be dropped....

Modern city planning is obliged to forego a significant number of artistic motifs. Regardless of how painful this may be to sensitive souls, the practical artist should not let himself be guided by sentimental impulses, because no artistic planning could be a thorough or lasting success unless it complied with modern living conditions. In our public life much has irrevocably changed, depriving certain old building forms of their original purpose, and about this nothing can be done. We cannot alter the fact that marketing has withdrawn more and more from the plazas, partly into inartistic commercial structures, partly to disappear completely because of direct delivery to the home. We cannot prevent the public fountains from being reduced to a merely ornamental role; the colorful, lively crowd stays away from them because modern plumbing carries the water much more conveniently directly into house and kitchen. Works of art are straying increasingly from streets and plazas in the "art-cages" of the museums; likewise, the colorful bustle of folk festivals, of carnivals and other parades, of religious processions, of theatrical performances in the open market place, etc., disappears. The life of the common people has for centuries been steadily withdrawing from public squares, and especially so in recent times. Owing to this, a substantial part of the erstwhile significance of squares has been lost, and it becomes quite understandable why the appreciation of beautiful plaza design has decreased so markedly among the broad mass of citizenry....

It is above all the enormous size to which our larger cities are growing that has shattered the framework of traditional artistic forms at every point. the larger the city, the bigger and wider the plazas and streets become, and the higher and bulkier are all structures, until their dimensions, what with their numerous floors and interminable rows of windows, can hardly be organized any more in an artistically effective manner. Everything tends toward the immense, and the constant repetition of identical motifs is enough to dull our senses to such an extent that only the most powerful effects can still make any impression. As this cannot be altered, the city planner must, like the architect, invent a scale appropriate for the modern city of millions.... Everywhere, as if spontaneously, lots are divided up and streets are broken through so that even in the old parts of town more and more side streets result, and something of the obnoxious building-block system surreptitiously takes over....

Passage 2
Beijing Looped Hybrid

An ultra-modern expression of 21st Century ecological urban living.

Filmic urban space; around, over and through multifaceted spatial layers, is one of the central aims of this 160,000 square meter Hybrid Building complex with over 700 apartments sited adjacent to the old city wall of Beijing.

Image courtesy Steven Holl Architects

Current development in Beijing is almost entirely äobject buildingsä and free standing towers. This äcity within a cityä envisions urban space as the central aim as well as all the activities and programs that can support the daily life of over 2500 inhabitants: café's, delis, laundry, dry cleaners, florists etc, line the main public passages.

The polychrome architecture of Ancient China inspires a new phenomenal dimension especially inscribing the äspatiality of the nightä. The undersides of the cantilevered portions are colored membranes in night light glow. Misting fountains from the water retention basin activate the night light in colorful clouds, while the floating Cineplex centerpiece has partial images of its ongoing films projected on its undersides and reflected in the water.

Image courtesy Steven Holl Architects

The eight towers are linked at the twentieth floor by a ring of cafes and services.

Focused on the experience of passage of the body through spaces, the towers are organized to take movement, timing and sequence into consideration. The point of view changes with a slight ramp up, a slow right turn. The elevator displaces like a ãjump cutä to another series of passages on a higher level, which pan across exhilarating peripheral views.

The encircled towers express a collective aspiration; rather than towers as isolated objects or private islands in an increasingly privatized city.... the hope of a new type of collective 21st. Century space in the air is inscribed.

Programmatically this loop aspires to be semi-lattice-like rather than simplistically linear. The hope is that the sky-loop and the base-loop will constantly generate random relationships, just as a modern city does.

Image courtesy Steven Holl Architects

Mass housing in China has historically been standardized and repetitive. To break the pattern this new vertical urban sector aspires to individuation in urban living with a huge variety of apartment lay-outs available among the 728 living spaces.

Digitally driven, prefabricated construction of the exterior structure of the eight towers allows for ãbeamlessä ceilings. Every apartment has two exposures with no interior hallways. Principles of Feng-Shui are followed throughout the complex, which is aimed at sustainability ãLEED Goldä rating.

Garden of Mounds, five landscape mounds with recreational activities, have been formed with the earth excavated from the new construction.

The new park is a semi-public space while the use of the integrated functions is electronically controlled by the resident's cards.
1. The Mound of Childhood is a fenced in area adjacent to and integrated with a kindergarten.
2. The Mound of Adolescence has a Basket Ball Court, a Roller Blade and Skate Board Area a Music and TV Lounge.
3. The Mound of Middle Age has a Coffee and Tea House, a Tai Chi Platform and two Tennis Courts.
4. The Mound of Old Age has Chess Tables, a Reading Lounge, a Tai Chi Platform and an Exer-

cise Machines Park.
5. The Mound of Infinity is a Meditation Place with ā5 Elementsä Pavilions: Earth, Wood, Metal, Fire and Water.

Site Area: 6.18 hectars
Total Building Area: 210,000 square meters

Construction start: November, 2004
Completion: Estimated 2006

Client:
Modern Hongyun Real Estate Development Co. LTD. Beijing, China

Architect:
Steven Holl Architects
Principal-in-charge: Steven Holl
Project Architect: Li Hu

Project Team:
Hideki Hirahara, Christiane Deptolla, Shih-I Chow, Matthew Uselman, Young Jang, Garrick Ambrose, Yenling Chen, James Macgillivray, Jongseo Lee, Judith Tse, Liang Zhao

Associated Architect:
Beijing Capital Engineering Architecture Design Co. LTD.

Structural Engineer:
Guy Nordenson and Associates

Mechanical Engineer:
TRANSSOLAR Energietechnik GmbH
Cosentini Associates

Lighting Designer:
Halie Light and L' Observatoire International

Resources for Reference

http://www.library.cornell.edu/Reps/DOCS/sitte.htm

http://www.arcspace.com/architects/Steven_Holl/beijing/index.htm

Section 3
Tips for Translation

汉译英中的词语省略

汉译英时应避免一字不漏的对译，显得生硬、累赘，而应当充分考虑英语的语言习惯和惯用语的使用，使译文简洁流畅。

1. ……力争建筑学专业领域对外交流的不断发展……
…in an effort to promote the foreign exchange in architecture…

本句中"建筑学专业领域"如果一定要"对等"翻译的话，译文就会出现"in the field of architecture specialty"，而实际上"architecture"一词本身就带有"专业领域"的含义。

2. 基地内将提供足够的机动车泊位和内部道路，以满足大会举办期间的使用要求，以及场馆平时使用的要求，例如训练。
Ample car-parking integrated with the internal circulation routes will be provided within the site to meet demands during events as well as non-event general uses of the facilities such as for training.

在本句中，应注意"机动车泊位"、"大会举办期间"等的翻译。

3. 我们在讨论九月份同哈佛大学联合设计的事宜。
We are discussing the joint studio with Harvard in September.
在汉语中，经常有"事宜"、"情况"、"现象"等说法，并没有明确和具体的含义，在英译时可予以省略。

Section 4
Tips for Writing

Be specific

On-line communication and the precision of the computer have little tolerance for the loose and the imprecise. An on-line search can be prolonged indefinitely by not being specific enough. But the need for precision isn't limited to the on-line message; it is a key ingredient of clear communication in any medium. Don't write "bring to reality" when you mean "build", say "partition", not "divider element". Avoid inexact space wasters such as "interesting", "impressive", "basically", and "situation"——as in "interview situation". They are filler words and, unless defined, add nothing to your message. Note that the use of a vague term where a specific one would work better often stems from vagueness of thought, and in such cases if the thought can be sharpened the words will come.

Consider the following sentences from a proposal introduction.
The self-contained instructional space——a splendid teaching medium for a specific objective——is simply inadequate for other tasks.
Work on the term "self-contained instructional space", then compare your answer with a solution below.
The enclosed classroom, suited to certain types of instruction, simply won't work for other tasks.

Unit 27

Section 1
Intensive Reading

Form-Resistant Structures
Mario Salvadori

Part IX

This chapter must end with the melancholy realization that over the last few years thin, curved shells, lovely as they may be, have not been very popular in advanced technological countries for purely economic reasons. The main obstacle to their popularity, already mentioned, is the cost of their curved formwork. Innumerable procedures have been invented and tried to reduce the cost of the formwork or to do away with it altogether. Pneumatic forms were first used in the 1940s by Wallace Neff, who sprayed concrete on them with a spraygun. Dante Bini sets the reinforcement and pours the concrete on uninflated plastic balloons, and then lifts them by air pressure.

The Bini procedure, in particular, has met with success almost all over the world in the erection of round domes of large diameter (up to 300 feet) for schools, gymnasiums, and halls. Of course, balloons are naturally efficient when round. These procedures cannot be well adapted to other thin-shell shapes.

A traditional method of construction, originating in the Catalonian region of Spain, has for centuries produced all kinds of curved thin structures without ever using complex scaffolds or formwork through the ingenious use of tiles and mortar. For example, to build a dome the Catalonians start by supporting its lowest and outermost ring of flat tiles on short, cantilevered, wooden brackets and grout to this first layer a second layer of tiles by means of a rapid-setting mortar. Once this first ring is completed and the mortar has set-in less than twelve hours-workers can erect the next ring by standing on the first and adding as many layers of tile as needed by the span of the dome, usually not more than three layers. By the same procedure, spiral staircases are erected around interior courtyards (Figure 27-1), or cylindrical barrels of groined vaults built. The Guastavino Company, whose Catalonian founder introduced this method to the United States toward the end of the nineteenth century, eventually built over 2,000 buildings in which such tile shells were used. Two of them, the dome over the crossing of the Cathedral of St. John the Divine (erected as a temporary structure while waiting for the completion of the church) and the groined vaults of the War College at Fort McNair, Virginia (Figure 27-2), have by now been officially labeled United States landmarks. Unfortunately, the amount of labor required to set the tiles by hand has made

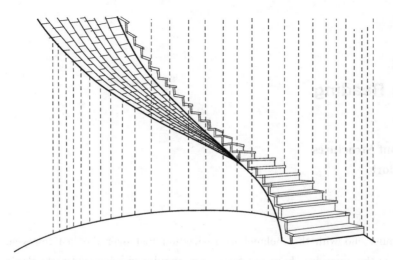

Figure 27-1 Spiral staircase of thin-shell tile construction.
(Saralinda Hooker and Christopher Ragus)

Figure 27-2 Groin vaults of thin-shell tile at Fort McNair, Virginia.
(Saraiinda Hooker and Christopher Ragus)

even this procedure uneconomical. The last word on this method's use has not been said, however, since in the [former] USSR thin shell specialists have extended the Catalonian methodology by replacing the small tiles with large, prefabricated, curved elements of prestressed concrete. These are erected without the need for any scaffold starting at one corner of a steel or concrete structural frame (Figure 27–3).

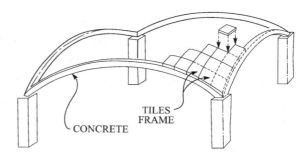

Figure 27–3 Thin shell of prefabricated elements. (Saralinda Hooker and Christopher Ragus)

In structures, perhaps more than in any other field of human invention, little is new under the sun, but there is always room for ingenious modifications of old ideas, as well as hope for real breakthroughs.

Words and phrases

1. melancholy   *adj.*   忧郁的
2. popularity   *n.*   普及，流行
3. pneumatic   *adj.*   充气的
4. spraygun   *n.*   喷枪/airbrush   喷笔
5. gymnasium   *n.*   体育馆/gym（缩写）/gymnastics   体操
6. rapid-setting mortar   快干砂浆
7. spiral staircase   螺旋楼梯
8. prefabricated   *adj.*   预制的
9. prestressed concrete   预应力混凝土
10. breakthrough   *n.*   突破

Section 2
Extensive Reading

Passage 1
City Planning According to Artistic Principles (Ⅴ)
Camillo Sitte
Translated by George R. Collins and Christiane Crasemann Collins (London: Phaidon Press, 1965): 91-104; 105-112

It would, moreover, be quite short-signed not to recognize the extraordinary achievements of modern city planning in contrast to that of old in the field of hygiene. In this our modern engineers, so much maligned because of their artistic blunders, have literally performed miracles and have rendered everlasting service to mankind. It is largely due to their work that the sanitary conditions of European cities have improved so remarkably——as is apparent from mortality figures which have in many cases been halved…. This we gladly grant, but there still remains the question as to whether it is really necessary to purchase their advantages at the tremendous price of abandoning all artistic beauty in the layout of cities.

The innate conflict between the picturesque and the practical cannot be eliminated merely by talking about it; it will always be present as something intrinsic to the very nature of things. This inner struggle between the two opposing demands is not, however characteristic of town planning alone; it is present in all the arts, even in those apparently the freest, if only as a conflict between their ideal goals and the limiting conditions of the material in which the work of art is supposed to take shape….

In the field of city planning the limitations on artistry of arrangement have, to be sure, narrowed greatly in our day. Today such a masterpiece of city planning as the Acropolis of Athens is simply unthinkable. That sort of thing is for us, at the moment, an impossibility. Even if the millions were provided that such a project would entail, we would still be unable to create something of the kind, because we lack both the artistic basis for it and any universally valid philosophy of life that has sufficient vigor in the soul of the people to find physical expression in the work. Yet even if the commission be devoid of content and merely decorative——as is the case with art today——it would be frightfully difficult for our realistic man of the nineteenth century. Today's city builder must, before all, acquire the noble virtue of an utmost humility, and, what is remarkable in this case, less for economic considerations than for really basic reasons.

Assuming that in any new development the cityscape must be made as splendid and pictorial as possible, if only decoratively in order to glorify the locality——such a purpose cannot be accomplished with the ruler or with our geometrically-straight street lines. In order to produce the effects of the old masters, their colors as well must form part of our palette. Sundry curves, twisted streets and irregularities would have to be included artificially in the plan; an affected artlessness, a purposeful unintentionalness. But can the accidents of history over the course of centuries be invented and constructed *ex novo* in the plan? Could one, then, truly and sincerely enjoy such a fabricated ingenuousness, such a studied naturalness? Certainly not the satisfaction of a spontaneous gaiety is denied to any cultural level in which building does not proceed at apparent random from day to day, but instead constructs its plans intellectually on the drawing board. This whole course of events, moreover, cannot be reversed, and consequently a large portion of the picturesque beauties we have mentioned will probably be irretrievably lost to use in contemporary planning. Modern living as well as modern building techniques no longer permit the faithful imitation of old townscapes, a fact which we cannot overlook without falling prey to barren fantasies. The exemplary creations of the old masters must remain alive with us in some other way than through slavish copying; only if we can determine in what the essentials of these creations consist, and if we can apply these meaningfully to modern conditions, will it be possible to harvest a new and flourishing crop from the apparently sterile soil.

An attempt should be made regardless of obstacles. Even if numerous pictorial beauties must be renounced and extensive consideration be given to the requirements of modern construction, hygiene, and transportation, this should not discourage us to the extent that we simply abandon artistic solutions and settle for purely technical ones, as in the building of a highway or the construction of a machine. the forever edifying impress of artistic perfection cannot be dispensed with in our busy everyday life. One must keep in mind that city planning in particular must allow full and complete participation to art, because it is this type of artistic endeavour, above all, that affects formatively every day and every hour the great mass of the population, whereas the theater and concerts are available only to the wealthier classes. Administrators of public works in cities should turn their attention to this matter.

Passage 2
CCTV——TV STATION AND HEADQUARTERS

This iconic new addition to the Beijing skyline combines the entire process of TV making——administration, production, broadcasting——into a single loop of interconnected activity. Two structures rise from a common production platform, each with a different character: one is dedicated to broadcasting, the second to services, research and education. They merge at the top to create a cantilevered headquarters for management.

Built directly adjacent to the CCTV, the Television Cultural Center (TVCC) will house public programs on site including a theater, cinemas, restaurants and five-star accommodation. The building will also serve as the international broadcasting center for the 2008 Olympics.

CCTV will be one among many towers in Beijing's new Central Business District, all striving to be unique-all different expressions of verticality.

Skyscraper
The tragedy of the skyscraper is that it marks a place as significant, which it then occupies and exhausts with banality…This banality is twofold: in spite of their potential to be incubators of new cultures, programs, and ways of life, most towers accommodate merely routine activity, arranged according to predictable patterns. Formally, their expressions of verticality have proven to stunt the imagination: as verticality soars, creativity crashes.

Concept
Instead of competing in the hopeless race for ultimate height——dominance of the skyline can only

be achieved for a short period of time, and soon another, even taller building will emerge——the project proposes an iconographic constellation of two high-rise structures that actively engage the city space: CCTV and TVCC.

CCTV combines administration and offices, news and broadcasting, program production and services——the entire process of TV-making——in a loop of interconnected activities. Two structures rise from a common production platform that is partly underground. Each has a different character: one is dedicated to broadcasting, the second to services, research and education; they join at the top to create a cantilevered penthouse for the management. A new icon is formed...not the predictable 2-dimensional tower "soaring" skyward, but a truly 3-dimensional experience. The consolidation of the TV program in a single building allows each worker to be permanently aware of the nature of the work of his co-workers——a chain of interdependence that promotes solidarity rather than isolation, collaboration instead of opposition. The building itself contributes to the coherence of the organization.

While CCTV is a secured building for staff and technology, public visitors will be admitted to the "loop", a dedicated path circulating through the building and connecting to all elements of the program and offering spectacular views across the multiple facades towards the CBD, Beijing, and the Forbidden City.

The Television Cultural Center (TVCC) is an open, inviting structure. It accommodates visitors and guests, and will be freely accessible to the public. On the ground floor, a continuous lobby provides access to the 1500-seat theater, a large ballroom, digital cinemas, recording studios and exhibition facilities. The building hosts the international broadcasting centre for the 2008 Olympic Games. The tower accommodates a five-star hotel; guests enter at a dedicated drop-off from the east of the building and ascend to the fifth floor housing the check-in as well as restaurants, lounges, and conference rooms. The hotel rooms are occupying both sides of the tower, forming a spectacular atrium above the landscape of public facilities.

On the block in the south-east, the Media Park is conceived as an extension of the proposed green axis of the CBD. It is open to the public for events and entertainment, and can be used for outdoor filming.

Resources for Reference

http://www.library.cornell.edu/Reps/DOCS/sitte.htm
http://www.oma.eu/index.php?option=com_projects&view=project&id=55&Itemid=10

Section 3
Tips for Translation

汉语英译多样化

  在汉译英中，为了避免译文单调，增加可读性，在保证翻译的准确性的同时，对同一种说法或同一个词可采取不同的译法。

  例如科技论中经常出现的"（文章的）作者认为……"可以译为"It is held in the article that…"，"It is proposed in the article that…"或者"The author holds that…"，甚至可以译为"It is pointed in the article that…"，等等。

  又如"本文从……出发"，可以译为"Starting from…, the article…"，或者"Taking…as its standpoint, the article…"

  再如"论文以……基础，……"则可译为"On the basis of…, the article…"或者"Base on…, the article…"。

  而常见的"调查研究"，则可结合具体的语境，翻译成"study"、"research into"、"probe"、"explore"、"investigate"等等。特别是在同一段文字出同样或者相似的汉语文字的情况下，更应注意英译的多样化。

  再来看下面几句句子：

1. 教育的职责是积极为国家培养各级专门人才。
It is the obligation of education to make vigorous effort to train people of various levels.

2. 学院近几年来积极推动对外学术交流与合作。
Our school is doing all what it can to promote the international academic communication and corporation.

3. 他意识到建筑的技术问题同样重要。
He recognized that the technical dimension of architecture is comparably important.

4. 我意识到自己犯了个错误。
I came to realize that I had made a mistake.

  无论是前两句中的"积极"一词，还是后两句中的"意识"一词，都有多种译法。

Section 4
Listening Practice

Please watch the video and answer the questions below.

1. Why was the sanitation garage built?

2. What is the major construction material of the salt shed and why the material was selected?
3. What are the functions of the façade?

Words and Phrases:

1. shed *n.* 棚，库
2. commission *v.* 委任
3. residential *n.* 住宅的
4. slope *n.* 坡度
5. cast-in-place *adj.* 现浇的
6. esprit de corps *n.* 团体精神
7. shield *n.* 盾
8. heat gain 热增量
9. rhythmic *adj.* 有韵律的，有节奏的；格调优美的
10. double facade 双层表皮
11. curtain wall 幕墙

Resource for Reference

https://www.youtube.com/watch?v=iUG1rtmAyxc

Unit 28

Section 1
Intensive Reading

INTERNATIONAL ARCHITECTURAL DESIGN COMPETITION
GUANGDONG OLYMPIC STADIUM

Part I

1.0 Design Synopsis

INTERNATIONAL ARCHITECTURAL DESIGN COMPETITION
GUANGDONG OLYMPIC STADIUM

THE PROPERTY OWNER:
GUANGDONG PHYSICAL EDUCATION AND SPORTS COMMISSION

THE COMPETITION ORGANIZER:
GUANGZHOU UREAN PLANNING BUREAU &
THE ORGANIZING COMMITTEE OF INTERNATIONAL
ARCHITECTURAL DESIGN COMPETITION FOR
GUANGDONG OLYMPIC STADIUM

1.1 Design References

The Guangdong Olympic Stadium will be the first phase of the major facilities amongst the proposed Guangdong Olympic Centre to be completed for the hosting of the 9th National Games of the People's Republic of China in Guangzhou in 2001, and for the staging of world-class sports events in future. The centre will cover a site of 975,873m^2 and includes new and redevelopment of existing facilities which at present consist of match fields for hockey, baseball, handball, shooting, archery and an equestrian centre, into a comprehensive international standard centre offering a main stadium seating 80,000 spectators, cycling tacks, a natatorium, tennis centre, sports-themed sculpture park, new centre, hi-tech exchange centre, athletes centre etc. .

The Olympic Stadium will occupy a site of 304,350m^2 and will be situated in the southern zone of the Olympic Centre. The 80,000-seat arena includes 600 seats at the rostrum, 300 seats for the media, 1,500 seats accommodated in boxes on the east and west wings, and provisions for the disabled. The stadium will be the venue for the opening ceremonies, soccer and track & field events of the 9th National Games. During non-competition periods the stadium will serve as a facility for sports training and the staging of national and international events such as celebration, exhibitions, concerts and other activities. The design combines the contemporary spirit with local characteristics, and aims to symbolize the energy of Guangzhou's social and economic building and aspirations to attain the status of a metropolis of vibrancy and creativity in the 21st century.

The design of the stadium conforms to the relevant building codes of the People's Republic of China, supplemented where appropriate by suitable codes and the most up-to-date experiences on design of similar facilities. An integrated approach will be adopted to address issues of fire-fighting, traffic planning, evacuation and crowd control, barrier-free design, seismic design, air-raid defense, sanitation and building automation etc. .

The principle references for design codes are:
    Fire Code for Architectural Design (GBJ 16—87)
    General Design Code for Civil Buildings (JGJ 37—87)

Words and phrases

1. synopsis *n.* 大纲
2. stage *vt.* 上演，举行
3. natatorium *n.* 游泳场，游泳池
4. rostrum *n.* 讲坛，演讲坛
5. disabled *n.* 残障者
6. metropolis *n.* 都会，大城市
7. conform to 符合，遵照
8. building code 建筑规范
9. evacuation *n.* 疏散，撤退
10. barrier-free design 无障碍设计
11. seismic design 抗震设计
12. air-raid defense 人防
13. sanitation *n.* 卫生，卫生设施

Section 2
Extensive Reading

Zoning:

Zoning shapes the city. Zoning determines the size and use of buildings, where they are located and, in large measure, the densities of the city's diverse neighborhoods. Along with the city's power to budget, tax, and condemn property, zoning is a key tool for carrying out planning policy. New York City has been a pioneer in the field of zoning policy since it enacted the nation's first comprehensive Zoning Resolution in 1916.

The New York City Zoning Resolution today is a blueprint for the development of the city. It is flexible enough to address the advances in technology, neighborhood transformations, emerging design philosophies and changing patterns of use that combine to make New York a great city.

The Zoning Resolution has two parts: zoning text and zoning maps. The text establishes zoning districts and sets forth regulations governing their land use and development. The maps show the locations and boundaries of the zoning districts.

New York City Zoning Districts

The city is divided into three basic zoning districts: residential (R), commercial (C), and manufacturing (M). The three basic districts are further divided into a variety of lower-density, medium-density and higher-density residential, commercial and manufacturing districts. Any of these districts may in turn be overlaid by special purpose zoning districts tailored to the unique characteristics of certain neighborhoods. Someblockfronts in residential districts may be overlaid as well by commercial districts providing for neighborhood retail stores and services. These overlay districts modify the controls of the underlying districts.

Each zoning district regulates:

√ permitted uses listed in one or more of 18use groups;
√ the size of the building in relation to the size of thezoning lot, known as the floor area ratio or FAR;
√ for residential uses, the number of dwelling units permitted, the amount of open space required on the zoning lot and the maximum amount of the lot that can be covered by a building (lot coverage);
√ the distance between the building and the front, side and rear lot lines;
√ the amount of parking required; and

√ other features applicable to specific residential, commercial or manufacturing districts.

Zoning Text

The zoning text consists of 12 articles. Articles I through VII contain the regulations that apply to every residential, commercial and manufacturing district. Articles VIII through XII contain the regulations for the special purpose districts.

Zoning Maps

The 126 New York City zoning maps show the locations and boundaries of zoning districts, special purpose districts and commercial overlay districts. Each map covers an area of approximately 8,000 feet (north/south) by 12,500 feet (east/west) and is printed at a scale of 1 inch = 1,200 feet.

Resources for Reference

http://www.nyc.gov/html/dcp/html/subcats/zoning.shtml
http://en.wikipedia.org

Section 3
Tips for Oral Presentation

Facial Expressions

Leave that deadpan expression to poker players. A speaker realizes that appropriate facial expressions are an important part of effective communication. In fact, facial expressions are often the key determinant of the meaning behind the message. People watch a speaker's face during a presentation. When you speak, your face-more clearly than any other part of your body-communicates to others your attitudes, feelings, and emotions.

Remove expressions that don't belong on your face.

Inappropriate expressions include distracting mannerisms or unconscious expressions not rooted in your feelings, attitudes and emotions. In much the same way that some speakers perform random, distracting gestures and body movements, nervous speakers often release excess energy and tension by unconsciously moving their facial muscles (e.g., licking lips, tightening the jaw).

One type of unconscious facial movement which is less apt to be read clearly by an audience is involuntary frowning. This type of frowning occurs when a speaker attempts to deliver a memorized speech. There are no rules governing the use of specific expressions. If you relax your inhibitions and allow yourself to respond naturally to your thoughts, attitudes and emotions, your facial expressions will be appropriate and will project sincerity, conviction, and credibility.

Resource for Reference

http://www.ljlseminars.com/facial.htm

Unit 29

Section 1
Intensive Reading

INTERNATIONAL ARCHITECTURAL DESIGN COMPETITION
GUANGDONG OLYMPIC STADIUM

Part II

1.2 Summary of Estimated Costs

Description	Total Cost (0,000RMB)	Cost Index	
		Quantity (m²)	Cost RMB/Unit
1. Builder's works and minor services			
1.1 Piling, foundation, basement structure	11000	70,000	1,571
1.2 Above grade structure	20000	70,000	2,857
1.3 PTFE coated fiberglass fabric and supporting frame	25000	29,000	8,621
1.4 External fac,ade	7000	47,000	1,489
1.5 Other builders'works and minor services	1300	70,000	186
Sub-total	64300	286,000	2,248,25
2. Mechanical and electrical installation			
2.1 Plumbing and drainage	500	70,000	71
2.2 Fire services	1500	70,000	214
2.3 Spectator stand lighting	2000	70,000	286
2.4 Electrical installation	4000	70,000	571
2.5 Public address and communication system	2000	70,000	286
2.6 Air-conditioning and mechanical ventilation system	3000	70,000	429
2.7 Lift installation	1000	70,000	143

续表

Description	Total Cost (RMB)	Cost Index	
		Quantity (m²)	Cost (RMB/Unit)
Sub-total	14000	70,000	2000
3. Interior decoration and fitting to public area	2000	–	–
4. Sport field surfacing, line marking and irrigation system	700	–	–
5. Sport specialist equipment comprising Score Board (2 NO.), clock system, camera system and wind speed indicator	8000	–	–
6. Signs, graphics, logo and feature	2000	–	–
7. East and West VIP links	4000	–	–
8. Contingencies	5000	–	–
Total	99370	70,000	14,195

1.3 Cost Summary

I The project costing in accordance with the Current Index for Project Cost Indexing in Guangdong Province has been completed and is appended, structural systems, building materials and construction technology are specified with due consideration of the prevailing national and local conditions to ensure the completion and commissioning of the stadium as scheduled.

II Scope of Cost Estimate

The cost estimate includes the following conventional construction costs:
1. Basement and Foundations
2. Structural frame and slabs
3. Finishing Works
4. Plumbing and Drainage
5. Fire Services Installations
6. Electrical works and general lighting
7. Ventilation system
8. Central air-conditioning to the VIP concourse and suites and the hotel
9. Jumbotron Score Board and Display Screens (2 sets)

The following are not included:
1. Interior decoration and fixtures for the sales/rental retail areas and the hotel.
2. Stage lighting and sound systems and satellite communication equipment for media centre.
3. Overheads such as operating cost, staff training, marketing and equipment.
4. Marketing costs
5. Site clearance and utilities diversion (if any)
6. Works outside site boundary, utilities and utility connection charges.
7. Finance charges and statutory charges
8. Consultants' fees
9. Import tax
10. Land premium

III This estimated cost summary serves only to provide a preliminary indication of cost and must be refined as the design develops.

IV The estimate is completed with Reference to the 2nd Market Price Index in Guangzhou, and has not been adjusted for price fluctuation when this estimate was being prepared, nor fluctuations in currency and foreign exchange rates.

Words and phrases

1. estimated cost 估算
2. piling *n.* 打桩,打桩工程
3. foundation *n.* 基础,地基
4. basement *n.* 地下室
5. grade *n.* 等级,坡度,斜坡,地表面与建筑物基础相遇的高度
6. coated *adj.* 涂上一层的
7. plumbing *n.* 管道工程
8. drainage *n.* 排水
9. public address and communication system 公共广播系统
10. ventilation *n.* 通风
11. slab *n.* 厚板
12. finishing *adj.* 最后的,完工的, *n.* 面层
13. concourse *n.* 汇合,集合,广场,(车站、机场)中央大厅
14. suites *n.* 套房
15. jumbotron *n.* (电视机的)超大屏幕
16. land premium 土地价格

Section 2
Extensive Reading

New York City Zoning
Residence Districts

Residence districts are the most common zoning districts in New York City, accounting for about 75 percent of the city's zoned land area. These districts accommodate an extraordinary variety of residential building forms——ranging from the single-family homes set amid wide lawns on the city's outskirts to the soaring towers of Manhattan.

To regulate such diversity, the Zoning Resolution designates 10 basic residence districts——R1 through R10. The numbers refer to permitted *bulk* and *density* (with R1 having the lowest density and R10 the highest) and other controls such as required parking. A second letter or number in some districts signifies additional controls. Unless otherwise stated, the regulations for each of the 10 residence districts pertain to all subcategories within that district. Regulations for the R4 district, for example, encompass R4-1, R4A and R4B districts, except where specific differences are noted.

Residences are also permitted in most commercial districts. Certain higher-density commercial districts mapped primarily in Manhattan are, in fact, substantially residential in character. In each commercial district, except C7 and C8, the size of a residential building or the residential portion of a *mixed building* is governed by the bulk provisions of a specified "equivalent" residential district. For example, R6 is the *residential district equivalent* of C4-2 and C4-3 districts.

All residence districts permit most *community facilities*, such as schools, houses of worship and medical facilities. In certain districts, in order to accommodate needed services, the maximum permitted *floor area ratio (FAR)* for community facilities exceeds the maximum permitted for residential uses. In districts limited to one- (family) and two-family homes, however, certain facilities are not permitted or are restricted in size.

LOWER DENSITY DISTRICTS (R1 – R5)

Lower-density residence districts are usually found far from central business districts in areas not well served by mass transit. These areas are characterized by low building heights, landscaped yards and high auto ownership. Some lower-density neighborhoods are comprised entirely of single-family *detached* homes on large lots, others have one- (family) and two-family detached homes on smaller lots, and still other neighborhoods have detached, *semi-detached* and *attached* buildings all mixed together (Figure 29–1).

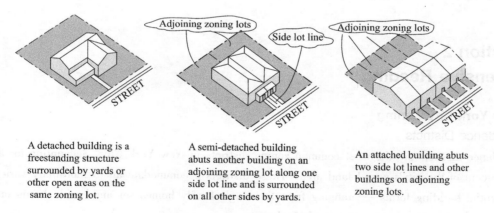

Figure 29–1 Lower density districts

Detached and semi-detached buildings typically accommodate either a single family or two families in separate *dwelling units*. Attached buildings may house one, two, or more families. R1 and R2 districts allow only detached *single-family residences*. R3A, R3X, R4A and R5A districts allow only detached single-family and *two-family residences*. R3-1 and R4-1 districts permit both detached and semi-family detached one- family and two-family houses. R4B districts also permit attached row houses limited to one and two families. R3-2, R4, R5, R5B and R5D are general residence districts that permit all housing types and are distinguished by differing bulk and density, height and *setback*, *lot coverage or open space*, and parking requirements.

Since 1989, R3, R4 and R5 districts with an A, B, D, X or 1 suffix have been created or revised to prevent the out-of-scale development that can blur distinctions between residence districts and alter the character of the city's traditional low-rise neighborhoods. The regulations for these new and revised districts aim to preserve neighborhood character by reaffirming the bulk distinctions, building configurations and narrow lot sizes of many residential neighborhoods. A maximum *building height* is established for each district, including the traditional low-rise row house districts (R4B and R5B). The familiar roof line of districts characterized by pitched roofs is encouraged by establishing a maximum *perimeter wall* height, above which pitched roofs or setbacks are required. In R2X, R3, R4, R4-1 and R4A districts, an increase in floor area is permitted for space beneath a pitched roof (*attic allowance*). The regulations promote landscaped front yards and allow for more on-street parking by controlling the location and dimensions of driveways and *curb cuts*.

Some R1, R2, R3, R4-1 and R4A districts in the city are designated as *Lower Density Growth Management Areas*, where residential developments are required to provide more parking spaces, larger yards and more open space. Designated areas include all such zoning districts in Staten Island and in Bronx Community District 10.

Resources for Reference

http://en.wikipedia.org/
http://www.nyc.gov/html/dcp/html/zone/zh_resdistricts.shtml
http://www.cityofchicago.org/Zoning/

Section 3
Tips for Translation

动词连用的英译

在汉语中经常出现两个以上的动词作谓语，有时是表示一种并列关系，有时是先后关系，但不用关联词连接。而在英语中并列的谓语动词一般必须通过关联词连接，并明确相互之间的关系。在汉译英中必须注意到这一点。

1. 规划师倾向于把不同的邻里各自为政分散布置，把一个区域划为居住而把另一个划为商务，或者在某些区域仅仅布置大学或文化设施用地。

City planners tend to lay out discretely different bounded neighborhoods, zone one area for housing and another for business, and establish certain areas just for universities or cultural facilities.

  本句的各个谓语动词就是一种相互并列的关系。前面根据汉语语序排列，用连词 and 结束最后一个部分。

2. 他站起身来，走到门口，把它扔了出去。
He rose up, walked to the door, and threw it out.

  本句的各个谓语动词是一种先后的关系。也用连词 and 带出最后一个谓语动词并结束整个句子。

  汉语中联用的谓语动词在有的时候往往包含目的、方式、结果等意思。在汉译英的时候，应尽可能选择恰当的英语表达方式，把这些含义传达出来。

3. 他去美国普林斯顿大学攻读建筑学硕士学位了。
He went to America to study for Master's Degree of Architecture in Princeton University.
  用不定式表示目的。

4. 建筑师修复了周边历史建筑，恢复了公共空间的活力。
The architect rehabilitated the public space by renovating the surrounding historical buildings.
  用介词段与表示方式。

5. 他引进新的建筑技术和材料，极大地减少了大量的能源消耗。
He introduced new building techniques and materials, greatly decreasing the energy consumption.
  用作状语的分词短语表示结果。

Unit 30

Section 1
Intensive Reading

INTERNATIONAL ARCHITECTURAL DESIGN COMPETITION
GUANGDONG OLYMPIC STADIUM

Part III

2.0　Master Layout

 2.1　The Site

I The site of the proposed Olympic centre is at the intersection of the Tianhe and Huangpu districts of Guangzhou, and is surrounded by low-density buildings and such established tourist and growth centre for cultural facilities such as the Amazing World, Wonders of Aviation theme parks, Science Centre and golf course. The site is immediately off the interchange between the eastern ring road to the west, and the Guangzhou-Shenzhen expressway to the north. The Shenzhen-Guangzhou railway, Zhongshan and Huangpu Avenues are to the south. The site is well connected to the centre of the capital and the other major cities in the Province.

II The stadium is located in the southern zone of the Olympic Centre. Connection to the other facilities of the centre will be at the north and north-west. The Stadium will be tied-in to the complex of the centre by external linkages on other sides. Vehicular access points are located on the west and south-east, and the main access is from the south. The site is basically flat with no existing buildings or structures.

 2.2　Site Planning

I The building is oriented on a north south axis to align with vehicular access and view corridors aligned with major architectural features of the stadium. There is a ring road that allows 360 degree circulation and free access to the stadium entries from various points. The loop road is wide enough to accommodate six lanes of traffic. There are multiple drop-off zones for the public.

II The entire stadium is recessed 8 meters from grade. This reduces the height of the stadium, enhancing its overall appearance. This allows the public to engage the main concourse level of the stadium. There are two main plaza areas (east and west) for gathering and celebration before during and after the event.

2.3 Traffic Planning

I The site is basically aligned north to south. In line with the established and planned traffic patterns, the connection to the transport networks are located at the western and southern periphery and includes 3 vehicular ingress and egress points, and the main access for spectators.

II The existing transport network consists of the main trunks of the Guangyuan Road, the Guangzhou-Shantou Highway, the Eastern City Ring Road expressway, Guangzhou-Shenzhen Railway and the district railway. The traffic-planning for the project concern the suitable interfacing with the extension of Guangyuan Road, Dongsha Road and the planned road to the east, branch-connection from the Guangzhou-Shenzhen Railway and the safe, efficient movements of upto 80,000 spectators.

III Ample car-parking integrated with the internal circulation routes will be provided within the site to meet demands during events as well as non-event general uses of the facilities such as for training.

IV The 3 vehicular access are located at the south connecting with Dongpo Road, at the northwest connecting with Dongsha Road, and via a circular road at the south-east connecting with the extension of Guangyuan Road. The south is the main spectator access which is served by a large public transport terminus. An independent access form Dongsha road is provided for the atheletes' and VIPs' security and convenience. Within the centre, oneway traffic will be maintained and pedestrians will be separated from vehicular traffic to minimize conflict, and to promote orderly and efficient road use. The centre complex will be designed to be barrier-free for the disabled.

V Circulation and Evacuation

Apart from a spectator capacity for 80,000 spectators, managers, coaches and other team members, security personnel, press and media and other ancillary staff, contribute to a highconcerntration of occupants in the stadium complex. Separate circulation and evacuation routes will need to be assigned to the different groups to ensure efficient circulation and safe evacuation of personnel in normal use as well as in exceptal circumstances.

Ⅵ Spectators Access

The building is fluid and breaks down the traditional approach of formal entry. Patrons are invited to experience the building in many different ways by many approaches. Turnstiles flank the plazas and are also located on the north and south. The entire main concourse is separated from the outside by means of a 3m tall gate system. These gates can be maneuvered in various configurations to provide multiple crowd control options for the proprietor.

Ⅶ VIP Access

There is a frontage road specified in the master plan. The design varies from the original concept to create a berm that raises the inner ring road, creating a special VIP entry on the cross-axis. This is a private and protected entrance for the world's most distinguished guests and state leaders. The opening is framed by a luxurious skylight that provides light into the private level and a magnificent view of the stadium before one descends. This leads to a private VIP drop-off underground. VIP′ then ascend a private elevator (compression) and arrive at the two levels of box suites overlooking the gaming field (release). These suites are adjacent to the VIP club lounges.

VIP and hotel guests can also arrive via the ring road. There are two core elements on the NW and SE. The entries are enhanced by a canopy that provides cover from the parking lane of the ring road to the entry vestibule. The VIP will allend the game enter through a turnstile and can take an escalator straight to the lounge or use one of four elevators that access all levels of the building. One elevator can be programmed for use by special VIP guests such as state leaders who would most likely attend some events.

Words and phrases

1. ring road   环路
2. expressway   *n.*   快速路
3. drop-off   下客
4. recess   *n.*   (墙壁等的) 凹进处,*vt.* 使凹进
5. ingress and egress   出入口
6. interface   *n.*   界面,分界面,接触面
7. capacity   *n.*   容量
8. turnstile   *n.*   十字转门,转栅
9. canopy   *n.*   (树) 冠,天篷,遮篷
10. vestibule   *n.*   门廊,前厅

Section 2
Extensive Reading

New York City Zoning
Zoning History

The ground-breaking **Zoning Resolution of 1916**, though a relatively simple document, established height and setback controls and designated residential districts that excluded what were seen as functionally incompatible uses. It fostered the iconic tall, slender towers that epitomize the city's prewar business districts and established the familiar context of three-story to six-story residential buildings found in much of the city.

While other cities were adopting the New York model, the model itself was constantly amended in response to major shifts in population and land use caused by a variety of factors: continuing waves of immigration; new mass transit routes and the growth corridors they created; the emergence of technology and consequent economic and lifestyle changes; the introduction of government housing and development programs; and, perhaps more than anything else, the rise of the automobile, which revolutionized land use patterns and created traffic and parking problems never imagined in 1916.

By mid-century, the 1916 document needed to be completely reconsidered. After lengthy study and public debate, the current Zoning Resolution was enacted and took effect in 1961.

The **1961 Zoning Resolution** coordinated use and bulk regulations, incorporated parking requirements and emphasized the creation of open space. It introduced incentive zoning by adding a bonus of extra floor space to encourage developers of office buildings and apartment towers to incorporate public plazas into their projects. In the city's business districts, it accommodated a new type of high-rise office building with large, open floors of a consistent size. Elsewhere in the city, the 1961 zoning dramatically reduced achievable residential densities, largely at the edges of the city (Figure 30-1).

Figure 30-1 Lever House, 1952

Time passes, urban design theories change, land uses

change, and zoning policy accommodates and anticipates those changes. Since passage of the 1961 ordinance, new approaches have been developed to deal with new challenges and new opportunities. A combination of incentive zoning, contextual zoning and special district techniques are being used to make zoning a more responsive and sensitive planning tool.

Resources for Reference

http://www.nyc.gov/html/dcp/html/zone/zonehis.shtml
http://geography.about.com/od/urbaneconomicgeography/a/zoning.htm

Section 3
Tips for Translation

汉译英中词语的选择和词类的转换

在汉译英的时候,要仔细体味汉语词汇的确切含义,结合英语的表达习惯或惯用法选择恰当的译法. 特别要避免不加甄别地从汉英词典中搬用英文翻译.
如下例:
民居: vernacular housing
民间组织: nongovernmental organization
民歌: folk songs

还有:
公共卫生: public health
卫生设施: sanitary facilities
卫生纸: toilet paper

再如:
环境保护: environmental protection
历史保护: historical preservation

在汉译英的时候,还必须注意的是,英译中的句子成分和词类词性没有必要同汉语对应,适当的词类的转换十分常见,以增加翻译的灵活性. 例如:

1. 项目的目的在于提高居住环境质量,重振城市中心区的活力.
The project aims at upgrading the living environment and revitalizing the city the center.
名词转译为动词.

2. 控制人口规模、减少能源消耗和保护生态环境是城市化过程中必须关注的.
Enough attention must be paid to population control, energy saving and environmental protection in urbanization.

　　动词转译为名词.

Unit 31

Section 1
Intensive Reading

INTERNATIONAL ARCHITECTURAL DESIGN COMPETITION
GUANGDONG OLYMPIC STADIUM

Part IV

3.0 Architectural Design

I The city of Guangzhou will usher in the new millennium with a "state of the art" stadium. The Guandong Olympic Stadium will stand as an architectural symbol of the "world's game". The design is envisioned to be unique and representative of one of the most wondrous and magical cultures, reflecting it's position in a global network of sport supporting nations. It is imbued with the energy of the world's most talented athletes, who will come to Guangzhou to compete at the highest levels of the game. The new facility will have all the amenities to make it a leader in the "high-tech" realm of modern sports. The building design readily adapts to the future. The building is a "stem of technology" that will continually grow along with the game and the Olympic Centre that will soon be an international mecca for sports activity in China.

II The building is fluid and breaks down the traditional approach of formal entry. Patrons are invited to experience the building in many different ways by many approaches. Turnstiles flank the plazas and are also located on the north and south. There are concession stands located around the perimeter of the bowl. These are major revenue generators and aid in the activation of the space. On the NW & SE corner are retail components including a team store. The team store will sell name-brand athletic merchandise and various team paraphernalia. These are accessible from the inside during game time and outside before and after.

Ⅲ   Once inside the stadium there are four triangular circulation ramps that lead from the main concourse to the upper deck. They are clad in a perforated metallic skin that captures the energy of people in motion while maintaining a sleek appearance.

Ⅳ   The Club lounges center around a grand triple height space. The lounges are activated by restaurants and bars. Club Lounge East and Club Lounge West will each include a Chinese restaurant. There will also be retail. The club lounges can generate revenue for the project when there is not an event. The main space can accommodate business meetings or serve as a party function space with views out to the Olympic Park. There are also opportunities for wedding Banquets.

Ⅴ   Overlooking Club Lounge West is a hotel. People visiting Guangdong Olympic Park or the city of Gungzhou (The stadium is a 10-minute cab ride from the center city) will be able to stay in the stadium's hotel yearlong. The entire program works in concert for both event and non-event functions. The hotel corridors look down onto the club lounges and help generate excitement for the overall project. People coming to watch the games may stay in the hotel and take advantage of shopping and other amenities that the Olympic Village will offer.

Ⅵ   Overlooking Club Lounge East are the team dormitories. Adjacent to the practice field, this area provides exciting accommodations for the worlds best athletes.

Ⅶ   The stadium roof covers 70% of the seating. The design team felt that a more open stadium design would greatly enhance the spectacle of the game. The ends of the stadium are open to let views and vistas become part of the event. The spectators, while enjoying the game, are also part of the Olympic Village experience. People outside the stadium are also attracted by being witness to the energy and excitement taking place during the game. Optionally, an alternative scheme allowing 100% of the seating to be covered has also been considered.

Ⅷ   The roof of the stadium is in honor of the athletes. It extends 75 meters at the midpoint via an enormous truss. It is a reference to those who push their bodies to extremes in the pursuit of excellence.

Ⅸ   Each scoreboard contains a jumbo-tron, and fixed advertising i. e. tri-vision panels and LED's for the communicating of information. Scoreboards also provide an exciting backdrop for the event, showcasing instant replay. The design integrates the traditional matrix board (generally used for score information) into a larger jumbotron. This electronic method of score-giving is integral with the main screen, similar to how figures are represented on television i. e. the sports segment of the nightly news. We find this provides a more flexible environment for the showcasing of game information/advertising and allows for a grander overall image during selected moments of the game. Screen technology is constantly evolving and there are many exciting products on the horizon, including laser beam technology that can project a crystal clear image form opposing sides of the field in sunlight conditions. This also allows for a non-conditioned zone behind the screen, creating an incredibly thin presentation board.

Words and phrases

1. envision   *vt.*   想象，预想
2. stand   *n.*   看台
3. paraphernalia   *n.*   随身用具
4. clad   *vt.*   在金属外覆以另一种金属
5. perforate   *vt.*   穿孔于
6. vista   *n.*   狭长的景色，街景
7. truss   *n.*   桁架

Section 2
Extensive Reading

NYC Zoning
GLOSSARY

This glossary provides brief explanations of planning and zoning terminology. Words and phrases followed by an asterisk are defined terms in the Zoning Resolution of the City of New York, primarily in Section. Consult the Zoning Resolution for the official and legally binding definitions of these words and phrases.

Air Rights (see Development Rights)

Development Rights

Development rights generally refer to the maximum amount of floor area permissible on a zoning lot. The difference between the maximum permitted floor area and actual floor area is referred to as "unused development rights." Unused development rights are often described as air rights.

A *zoning lot merger* is the joining of two or more adjacent zoning lots into one new zoning lot. Unused development rights may be shifted from one lot to another, as-of-right, only through a zoning lot merger (Figure 31-1).

A *transfer of development rights* (*TDR*) allows for the transfer of unused development rights from one zoning lot to another in special circumstances, usually to promote the preservation of historic buildings, open space or unique cultural resources. For such purposes, a TDR is permitted where the transfer could not be accomplished through a zoning lot merger because certain conditions, such as intervening streets, separate the zoning lots. In the case of a landmark building, for example, a transfer may be made by CPC special permit from the zoning lot containing the designated landmark to an adjacent zoning lot or one that is directly across a street or, if the landmark is on a corner lot, diagonally across an intersection (Figure 31-2).

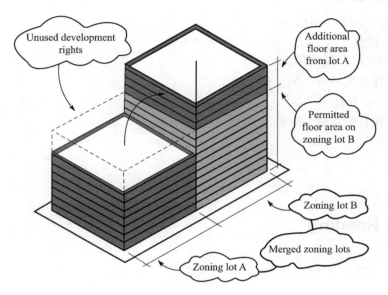

Figure 31-1

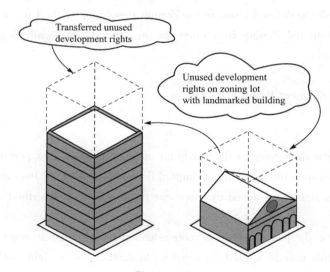

Figure 31-2

Floor Area

The floor area of a building is the sum of the gross area of each floor of the building, excluding mechanical space, cellar space, floor space in open balconies, elevators or stair bulkheads and, in most zoning districts, floor space used for accessory parking that is located less than 23 feet above curb level.

Floor Area Ratio (*FAR*)

The floor area ratio (FAR) is the principal bulk regulation controlling the size of buildings. FAR is the ratio of total building floor area to the area of its zoning lot. Each zoning district has an FAR control which, when multiplied by the lot area of the zoning lot, produces the maximum amount of floor area allowable in a building on the zoning lot. For example, on a 10,000 square-foot zoning lot in a district with a maximum FAR of 1.0, the floor area of a building cannot exceed 10,000 square feet (Figure 31-3).

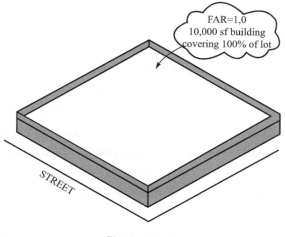

Figure 31-3

Resources for reference

http://en.wikipedia.org
http://www.nyc.gov/html/dcp/html/zone/glossary.shtml
http://www.fpza.org/index.shtml

Section 3
Tips for Translation

汉译英句子中的处理—断句

　　在翻译汉语中的长句的时候，应根据句子结构和意思的层次，适当断句，不必硬译成一句复杂冗长的长句。

1. 东方明珠电视塔位于黄浦江畔，浦东陆家嘴尖上，塔高468米，三面环水，与外滩的万国建筑博览群隔江相望，是亚洲第一、世界第三的高塔。
The Oriental TV Tower stands on the Hunagpu River and at the point of Lujiazui in Pudong. Surrounded by waters on three sides and facing the row of western-style architectures in the Bund area across the river, the 468-meter-tall tower ranks first in Asia and third in the world.
　　本句翻译划分为两个部分，前一部分表达东方明珠电视塔的位置，后一部分表达其高度和排名。

2. 广东奥林匹克体育场占地975873平方米，是规划中的广东奥林匹克中心主场馆的一期工程。

The Guangdong Olympic Stadium will cover a site of 975,873m². It is the first phase of the major facilities amongst the proposed Guangdong Olympic Centre

3. 在体育场内部有四条从主要的集散空间通向上层看台的三角形的坡道，外部覆盖穿孔金属表皮，以搜集活动人群散发的能量，同时保持光滑的外表。
Once inside the stadium there are four triangular circulation ramps that lead from the main concourse to the upper deck. They are clad in a perforated metallic skin that captures the energy of people in motion while maintaining a sleek appearance.

4. 俯瞰东侧休息区的是宿舍部分，靠近训练场，能为世界上最出色的运动员提供满意的食宿条件。
Overlooking Club Lounge East are the team dormitories. Adjacent to the practice field, this area provides exciting accommodations for the worlds best athletes.

5. 世界范围内体育场业主（主要是政府部门）对公共集会设施经济性的日益重视导致了建设目标从社会服务转向对经济可行性和设施运行使用最优化的关注。
Stadium owners, primarily government bodies, around the world have become more conscious of the financial performance of the public assembly facilities they own. This has resulted in a change in the objectives from community service to a concern regarding financial accountability and optimization of use and operating revenues.

Unit 32

Section 1
Intensive Reading

INTERNATIONAL ARCHITECTURAL DESIGN COMPETITION
GUANGDONG OLYMPIC STADIUM

Part V

4.0 Business Operation of Stadium

I   Stadium owners, primarily government bodies, around the world have become more conscious of the financial performance of the public assembly facilities they own. This has resulted in a change in the objectives from management for community service and public profile to a concern regarding financial accountability and optimization of use and operating revenues. This has resulted in the need to place greater emphasis on income-generating elements and the needs of the varied users of the facility.

II   The increasing sophistication of the facilities and the resulting business operations also require a more complex and sophisticated management approach. Stadium operations is now a multi-million dollar business requiring the management of assets, which can be valued at, millions of US dollars with a similar annual turnover.

III   The incorporation of a strong retail mix which is strategically planned and cohesively executed is vital to the financial success of the Stadium on a long term basis. The individual components of the retail offering will be structured so that they collectively provide a diverse range of activities which will offer all the facilities for an active day out:

- Local Restaurants
- International Restaurants, e. g. Hard Rock Cafe
- Fast Food Outlets, both local and international
- International Health Club
- Gymnasium
- Sports Museum
- Sports & Entertainment Bar
- Sports and Related Theme Shops
- Children's Amusement

These offerings will become important support aspects in the life of the Stadium during a major event and will act as the primary focus between such events.

Ⅳ Through thematic development and strategic planning of the retail offering to support the overall brand, the Stadium will be uniquely positioned in the region to draw major events. Following a complete analysis and needs assessment, a comprehensive strategy will be developed to determine the precise retail mix required to attract a maximum number of users, both during and between major events. Brand equity in the Stadium will, therefore, be enhanced through associations with key local, national and international brands.

5.0 Identity and Image

Ⅰ In order to effectively promote and market the Stadium as a competitive venue, a unique and proprietary identity will need to be developed as a Brand for the Stadium. This brand would comprise a unique name, a proprietary graphic symbol and representative colors, which when coupled with a solid branding strategy, would enable the Stadium to be strategically positioned and marketed against its global competitors. A strong brand for the Stadium will quickly establish recognition and value which will consequently provide unique promotional opportunities that can generate revenue as well as build broader recognition of the Stadium and its events. This brand will principally be used on and within the facility itself and will be the primary identity of the Stadium for the longer term.

Ⅱ One of the primary objectives in the creation of the identity is to integrate the graphic expression with the architecture so that the entire experience and visual language of the Stadium in unified and cohesive. This link between the two elements will provide the logic for the graphic design and will strongly reinforce the identity of the Stadium. This is an important connection in that the Stadium is not merely reflecting the local sporting team, rather it is a multi-purpose complex capable of staging diverse events from football and other athletic events to music and performances. Thus the brand needs to be associated with the venue rather than any particular team or event.

III   A full brand identity structure would also be developed in support of the basic identity in order to create a comprehensive and consistent communication system. This system would reflect the key attributes of the Stadium, differentiate it from the competition and be used throughout the complete promotional system as an identifier of the Stadium and its endorsed events and offerings.

IV   In terms of communications, the identity will be expanded to accommodate all the needs of advertising, promotions and public relations, as well as external and internal publications.

V   Pertaining to facilities, the identity will be translated to a proprietary signage system, uniforms, vehicles, merchandise and other applications related to the use and promotion of the facilities.

VI   In support of associated offerings, sub-brands will be developed for proprietary events and retail outlets developed by the Stadium. These, in turn, will be expanded to respond to the respective communication and facility needs for each sub-branded offering.

6.0 Signage

I   Signage plays many different roles in a development of this scale. The primary function of signage for the Stadium is to herald the identity of the complex and in so doing, it reinforces the graphic image which represents the Brand of the Stadium. This brand, in turn, can then be applied on permanent and temporary signage to endorse the various types of events which will be held within it.

II   The secondary functions of signage include directional and informational guidance. Wayfinding schemes will be developed in order to ensure that the 80,000 spectators can logically and easily maneuver through and around the Stadium. It is of particular importance in a development of this scale that the users be able to quickly identify, at every point of decision-making, where they are, what their choices are and how they can most efficiently get from one point to the next.

III   The tertiary functions of signage are mainly regulatory. In addition to aiding circulation and decision-making, signage must also ensure that life-safety is facilitated. Through legible and well-placed signage, means of escape can be quickly and easily identified in order to ensure the greatest possible safety of spectators and employees alike.

IV   The design of the signage should reflect the architecture and be both visually and tectonically sympathetic with the detailing of the Stadium. As a natural extension of the building it will act in a supporting role to reinforce the logic of the Stadium's design and will thus enhance the overall identity.

Words and phrases

1. optimization   *n.*   最佳化，最优化
2. turnover   *n.*   营业额
3. thematic   *adj.*   主题的
4. strategic   *adj.*   战略的，战略上的
5. attribute   *n.*   属性，品质，特征
6. proprietary   *adj.*   所有的，私人拥有的
7. signage   *n.*   标识

Section 2
Extensive Reading

NYC Zoning
GLOSSARY
Block
A block is a tract of land bounded on all sides by streets or by a combination of streets, public parks, railroad rights-of-way, pierhead lines or airport boundaries.

Mixed Building
A mixed building is a building in a commercial district used partly for residential use and partly for community facility or commercial use. A building that contains any combination of uses is often referred to as a mixed-use building. When a building contains more than one use, the maximum FAR permitted on the zoning lot is the highest FAR allowed for any of the uses, provided that the FAR for each use does not exceed the maximum FAR permitted for that use. In a C1-8A district, for example, where the maximum commercial FAR is 2.0 and the maximum residential FAR is 7.52, the total permitted FAR for a mixed residential/commercial building would be 7.52, of which no more than 2.0 FAR may be applied to the commercial space.

Mixed Use District
A mixed use district is a special zoning district in which new residential and non-residential (commercial, community facility and light industrial) uses are permitted as-of-right. In these districts, designated on zoning maps as MX with a numerical suffix, an M1 district is paired with an R3 through R9 district.

Open Space
Open space is the part of a residential zoning lot (which may include courts or yards) that is open

and unobstructed from its lowest level to the sky, except for specific permitted obstructions, and accessible to and usable by all persons occupying dwelling units on the zoning lot. Depending upon the district, the amount of required open space is determined by the open space ratio, minimum yard regulations or by maximum lot coverage.

Setback

A setback is the portion of a building that is set back above the base height (or street wall or perimeter wall) before the total height of the building is achieved. The position of a building setback in height factor districts is controlled by sky exposure planes and, in contextual districts, by specified distances from street walls.

Resources for Reference

http://en.wikipedia.org
http://www.nyc.gov/html/dcp/html/zone/glossary.shtml
http://www.ci.cambridge.ma.us/cdd/cp/zng/zord/index.html

Section 3
Tips for Oral Presentation

Impromptu or Extemporaneous Speaking

While many of us do not like to speak in front of people, there are times when we are asked to get up and say a few words about someone or a topic when we have not planned on saying anything at all. We are more shocked than anyone else. Has this ever happened to you? If and when this does happen to you, be prepared to rise to the challenge. Below are some tips you can use the next time you are called on to speak.

- Decide quickly what your one message will be——Keep in mind you have not been asked to give a speech but to make some impromptu remarks. Hopefully they have asked you early enough so you can at least jot down a few notes before you speak. If not, pick ONE message or comment and focus on that one main idea. Many times, other ideas may come to you after you start speaking. If this happens, go with the flow and trust your instincts.
- Do not try and memorize what you will say——Trying to memorize will only make you more nervous and you will find yourself thinking more about the words and not about the message.
- Start off strong and with confidence——If you at least plan your opening statement, this will get you started on the right foot. After all, just like with any formal speech, getting started is the most difficult. Plan what your first sentence will be. You may even write this opening line down

on your note card and glance at it one more time just before you begin speaking. If you know you have three points or ideas to say, just start off simple by saying, "I would just like to talk about 3 points". The first point is…the second point is…and so on.

- Decide on your transitions from one point to the other——After you have decided on your opening remark or line, come up with a simple transition statement that takes you to your main point. If you have more than one point to make, you can use a natural transition such as, "My second point is…" or "my next point is…" etc. Just list on your note card or napkin, if you have to, the main points or ideas. Do not write out the exact words, but just the points you want to mention.

- Maintain eye contact with the audience——This is easier to do if you do not write down all kinds of stuff to read. Look down at your next idea or thought and maintain eye contact with your audience and speak from your heart. Focus on communicating TO your audience and not speaking AT the crowd.

- Occasionally throw in an off-the-cuff remark——Because you want your style to be flexible and seem impromptu, trust your instinct and add a few words which just pop into your head. Keep it conversational and think of the audience as a group of your friends.

- Finally, have a good conclusion——Gracefully just state, "And the last point I would like to make is …." Once you have made your last point, you can then turn control back to the person who asked you to speak in the first place.

With a little practice, this process will feel more natural to you. Anticipating that you MAY be asked to say a few words should force you to at least think about what you might say if you are asked. Then if you ARE asked, you are better prepared because you anticipated being asked. This is much better than thinking they won't ask you and they actually do!

Resource for Reference

http://www.ljlseminars.com/imprompt.htm

Unit 33

Section 1
Intensive Reading

"Im Birch" School

Part I

Zentrum Zurich Nord

The restructuring and relocation of production for the industries based in Oerlikon marked the starting point for rapid changes to an inter-city area measuring some 60 hectares in size. The existing, large-scale manufacturing buildings and their development pattern, along with the siting of four different open, recreational areas——part of the overall planning concept——defined the formal structure for ongoing urban redevelopment. These guidelines were the result of an urban planning competition held in 1992 for the Zentrum Zurich Nord, a new city district designed to provide homes for 5,000 and jobs for 12,000.

The "Im Birch" School, situated on the northern boundary of the area covered by the plans, is the largest school complex in Zurich. It provides facilities for 700 pupils within two predefined building complexes, each with a stipulated maximum building height. The magnitude and complexity of the use requirements, the result of combining several stages of education under one roof (nursery, primary school and secondary school, plus after-school care facilities, common areas and sports hall), placed high demands on the layout of the school. At the same time, the design had to be flexible enough to take account of future requirements while allowing for the needs of current teaching methods (Figure 33–1、Figure 33–2).

Situation

Peter Markli placed two buildings on the plots, one rather flatter and elongated, the other more

Figure 33-1 Photo of model Zentrum Zurich Nord, after completion, 1:1000, view from south

Figure 33-2 General view of school buildings

compact and taller. The relationship between the two buildings is quite definite thanks to their positioning and the volumetric "subtractions"; they are seen as one coherent, sculpted figure. The more northerly building is divided into two distinct parts: the sports hall and a four-storey wing. The latter houses the primary school and the common facilities such as a multi-purpose hall, library and dining hall.

The four-storey building (with a ceiling height of 3.5m in contrast to the 3.0m of the northern building) on the southern plot contains the secondary school and the nursery, and together with the covered bicycle racks marks the southern limit of the development. The forecourt forms part of the overall plan for the open areas and also serves as a link between Oerliker Park and Friedrich Traugott Wahlen Park to the east. Large-format in situ concrete slabs create a coherent paved area which——due to the choice of materials and the form——stands out clearly from the neighbouring paths and roads, positioning the school complex as a distinctive public facility in the Zentrum Zurich Nord.

Together with the adjoining developments and the parks, the volumetric arrangement of the complex defines external areas with changing boundaries. This is a strategy that helps provide the individual levels of education within this large complex with their own access zones and their own external areas. At the same time, it helps to integrate this group of buildings into its environment (Figure 33-3 ~ Figure 33-14).

Figure 33-3 Model, 1:500

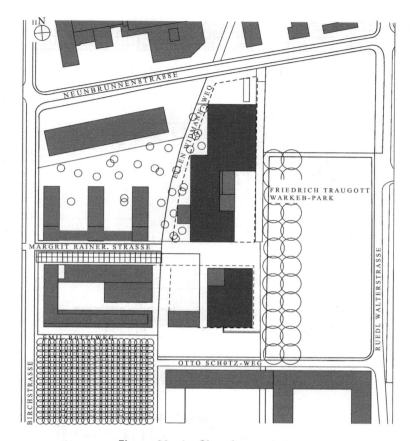

Figure 33–4 Site plan, 1:3000

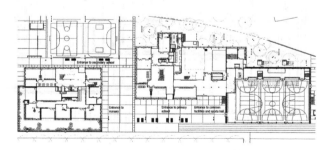

Figure 33–5 Plan of ground floor

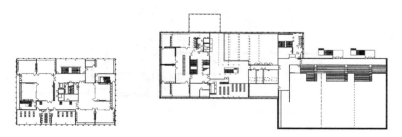

Figure 33–6 Plan of 1st floor

Figure 33-7　East elevation

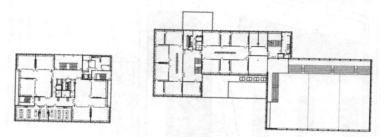

Figure 33-8　Plan of 2nd floor

Figure 33-9　Plan of 3rd floor

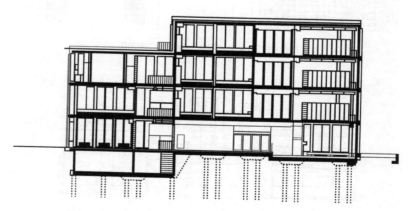

Figure 33-10　Section through secondary school

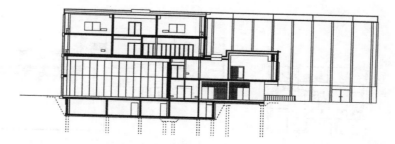

Figure 33-11　Section through primary school and common facilities

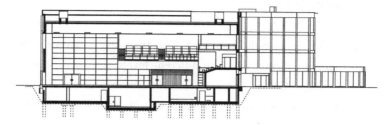

Figure 33-12   Section through sports hall

Figure 33-13   Lobby in teaching unit of secondary school

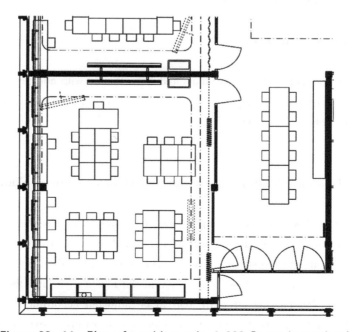

Figure 33-14   Plan of teaching unit, 1:200 Secondary school

Internal layout and classification

The idea of allocating certain areas to the individual education levels externally is continued inside the building. There are groups of rooms for the different levels and these form independent units within the parts of the building. Groups of two to four classrooms plus one or two group rooms, together with a common area, form one teaching unit, a sort of small school within the larger establishment.

The proposed internal layout with the hall bounded by classrooms on three sides makes for a building with a significant depth. In order to provide adequate lighting for the central areas, the walls of the hall are glazed for the full height of the room. This transparency and the arrangement of the rooms enables clearly structured, interdisciplinary teaching and, by including the shared hall, various other different teaching methods as well. Curtains are used to regulate the views into the individual classrooms.

This layout, characterised by the central hall or the lobby, differentiates this school from conventional ones, where the classrooms are usually reached via a system of corridors. Identifiable places have been created within the school complex at the level of the individual teaching stages to reinforce the pupils' identification with the school. Another crucial aspect of this layout is the "deconcentration system", which was required by the local fire brigade. What this entails is a third exit for all classrooms to guarantee an escape route that does not pass through the hall; that enables the hall to be furnished without any restrictions.

Design and loadbearing structure

The loadbearing structure is a system of columns and slabs braced with additional fair-face concrete walls to resist horizontal forces. Lightweight elements, bricks and glass block walls constitute the non-loadbearing elements (Figure 33-15).

Isometric view
Checking the room layout and loadbearing structure around the music room, dining hall and library (sketch)
The rational facade layout with its projecting lesenes (pilaster strips) seems to indicate a corresponding arrangement of the loadbearing columns behind. However, a closer look reveals variations in the column layout and the structural walls. The placement of the teaching units and the interlacing of different structures and room sizes, around the music room and the sports hall for instance, meant that the loadbearing structure had to be adjusted accordingly. In particular, the structure at ground floor level was determined by references to external spaces and the position of entrances. Around the entrance to the common facilities and the sports hall, as well as the covered

external facilities for the nursery, the loads from the columns above are carried on downstand beams acting as transfer structures (Figure 33 – 16、Figure 33 – 17).

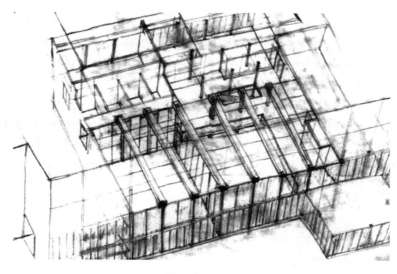

Figure 33 – 15

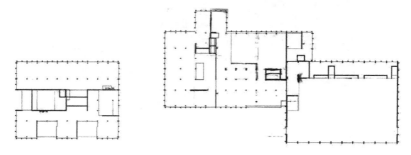

Figure 33 – 16   Loadbearing structure, ground floor, 1:1000
Plans of design process with additions by hand

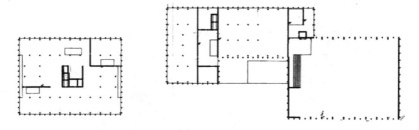

Figure 33 – 17   Loadbearing structure, 2nd and 3rd floors, 1:1000

The floor slabs are 340mm thick in order to bridge the long spans in some areas. But even where the spans are shorter the same thickness is used for economic reasons. This great mass of concrete renders impact sound insulation unnecessary.

The dark colouring of the prefabricated, slender (250×250mm) fair-face concrete columns is due to the properties of the aggregates used and the high proportion of cement. This high-strength concrete complies with enhanced structural requirements in terms of the compressive strength. The loads are transferred to the subsoil via a concrete pile foundation, with piles up to 27m long.

The surface finish to the structural fair-face concrete walls is achieved by using formwork type 2, i.e. a uniform surface texture is achieved without specifying the size of formwork panel to be used, which depends on the formwork system employed. Only the direction of the joints between panels was specified by the architects; their position and appearance was then decided by the contractor (Figure 33-18 ~ Figure 33-20).

Figure 33-18

Figure 33-19

Figure 33-20

Words and phrases

1. hectares   *n.*   公顷（hectare 的复数）
2. recreational   *adj.*   娱乐的，消遣的；休养的
3. complex   *n.*   复合体；综合设施
4. contractor   *n.*   承包人
5. northerly   *adj.*   北方的，向北的；来自北方的   *n.*   北风   *adv.*   向北；来自北方
6. bicycle rack   自行车托架
7. forecourt   *n.*   前院；前场
8. in situ   现浇
9. site plan   总设计图；总平面图；修建性详细规划
10. elevation   *n.*   立面图
11. section   *n.*   截面；
12. glazed   *adj.*   光滑的；像玻璃的；上过釉的；   *v.*   上釉（glaze 的过去分词）；装以玻璃
13. deconcentration   *n.*   分散
14. isometric   *adj.*   等距的；等角的；等轴的
15. non-loadbearing   非承重的
16. pilaster   壁柱，半露柱
17. pile foundation   桩基础
18. formwork   *n.*   量规，模架；模板
19. downstand beam   肋形楼板梁
20. transfer structures   转换层结构
21. prefabricated   *adj.*   预制构件的   *v.*   预先建造组合（prefabricate 的过去分词）
22. facade   *n.*   正面；表面；外观
23. precast   *adj.*   预制的   *vt.*   预制；预浇制
24. plinth   *n.*   柱基；底座
25. rendered   *v.*   提出；描绘（render 的过去分词）；放弃；报答；   *adj.*   已渲染的
26. substrate   *n.*   基质；基片；底层（等于 substratum）
27. drainage mat   排水砂垫层
28. separating layer   分离层，分层，分隔层
29. mineral-fibre   矿物纤维
30. vapour barrier   隔汽层
31. adhesive   *n.*   粘合剂；胶黏剂   *adj.*   黏着的；带黏性的
32. linoleum   *n.*   油布；油毡；漆布
33. screed   *n.*   ［地质］砂浆层
34. perforated   *adj.*   穿孔的；有排孔的   *v.*   穿孔（perforate 的过去分词）
35. gypsum board   石膏板
36. luminaires   *n.*   灯具（luminaire 复数）；照明器

37. blinding layer   草鞋底；地基垫层
38. lean concrete   贫混凝土，少灰混凝土

Section 2
Extensive Reading

Heating

Part I

A perfectly sealed and insulated building would hold heat for ever and thus would need no heating. The two dominant reasons why buildings lose heat are:
1. Conduction-heat flowing directly through walls, windows and doors;
2. Ventilation-hot air trickling out through cracks, gaps, or deliberate ventilation ducts.
In the standard model for heat loss, both these heat flows are proportional to the temperature difference between the air inside and outside. For a typical British house, conduction is the bigger of the two losses, as we'll see.

Figure 33-21

kitchen	2
bathroom	2
lounge	1
bedroom	0.5

Table 33-1. Air changes per hour: typical values of N for draught-proofed rooms. The worst draughty rooms might have N = 3 air changes per hour. The recommended minimum rate of air exchange is between 0.5 and 1.0 air changes per hour, providing adequate fresh air for human health, for safe combustion of fuels and to prevent damage to the building fabric from excess moisture in the air (EST 2003).

Conduction loss

The rate of conduction of heat through a wall, ceiling, floor, or window is the product of three things: the area of the wall, a measure of conductivity of the wall known in the trade as the "U-value" or thermal transmittance, and the temperature difference-

$$\text{power loss} = \text{area} \times U \times \text{temperature difference}.$$

The U-value is usually measured in $W/m^2/K$. (One kelvin (1 K) is the same as one degree Celsius (1°C).) Bigger U-values mean bigger losses of power. The thicker a wall is, the smaller its U-value. Double-glazing is about as good as a solid brick wall. (See table 33-2.)

U-values of walls, floors, roofs, and windows.   Table 33-2

	U-values ($W/m^2/K$)		
	old buildings	modern standards	best methods
Walls		0.45—0.6	0.12
solid masonry wall	2.4		
outer wall: 9 inch solid brick	2.2		
11 in brick-block cavity wall, unfilled	1.0		
11 in brick-block cavity wall, insulated	0.6		
Floors		0.45	0.14
suspended timber floor	0.7		
solid concrete floor	0.8		
Roofs		0.25	0.12
flat roof with 25mm insulation	0.9		
pitched roof with 100mm insulation	0.3		
Windows			1.5
single-glazed	5.0		
double-glazed	2.9		
double-glazed, 20mm gap	1.7		
triple-glazed	0.7—0.9		

The U-values of objects that are "in series," such as a wall and its inner lining, can be combined in the same way that electrical conductances combine:

$$u_{\text{series combination}} = 1 \Big/ \left(\frac{1}{u_1} + \frac{1}{u_2} \right)$$

There's a worked example using this rule on page 296.

Ventilation loss

To work out the heat required to warm up incoming cold air, we need the heat capacity of air: $1.2 kJ/m^3/K$.

In the building trade, it's conventional to describe the power-losses caused by ventilation of a space as the product of the number of changes N of the air per hour, the volume V of the space in cubic metres, the heat capacity C, and the temperature difference ΔT between the inside and outside of the building

$$\text{power(Watts)} = C\frac{N}{1\text{h}}V(\text{m}^3)\Delta T(K)$$
$$= (1.2\text{KJ/m}^3/K)\frac{N}{3600}V(\text{m}^3)\Delta T(K)$$
$$= \frac{1}{3}NV\Delta T$$

Energy loss and temperature demand (degree-days)

Since energy is power × time, you can write the energy lost by conduction through an area in a short duration as

$$\text{energy loss} = \text{area} \times U \times (\Delta T \times \text{duration}),$$

and the energy lost by ventilation as

$$\frac{1}{3}NV \times (\Delta T \times \text{duration}).$$

Both these energy losses have the form

$$\text{Something} \times (\Delta T \times \text{duration}),$$

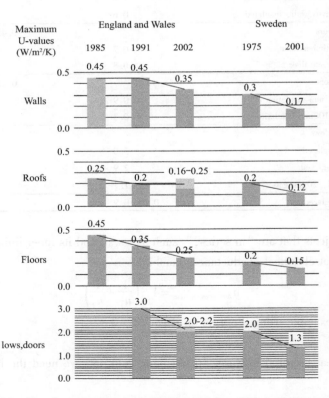

Figure 33-22 U-values required by British and Swedish building regulations.

where the "Something" is measured in watts per°C. As day turns to night, and seasons pass, the temperature difference ΔT changes; we can think of a long period as being chopped into lots of small durations, during each of which the temperature difference is roughly constant. From duration to duration, the temperature difference changes, but the Somethings don't change. When predicting a space's total energy loss due to conduction and ventilation over a long period we thus need to multiply two things:

1. the sum of all the Somethings (adding area × U for all walls, roofs, floors, doors, and windows, and $\frac{1}{3}$NV for the volume); and

2. the sum of all the Temperature difference × duration factors (for all the durations).

The first factor is a property of the building measured in watts per ℃. I'll call this the leakiness of the building. (This leakiness is sometimes called the building's heat-loss coefficient.) The second factor is a property of the weather; it's often expressed as a number of "degree-days," since temperature difference is measured in degrees, and days are a convenient unit for thinking about durations. For example, if your house interior is at 18℃, and the outside temperature is 8℃ for a week, then we say that that week contributed 10 × 7 = 70 degree-days to the (ΔT × duration) sum. I'll call the sum of all the (ΔT × duration) factors the temperature demand of a period.

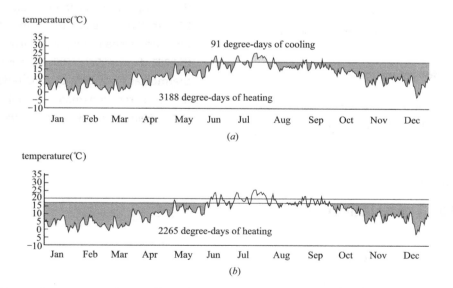

Figure 33-23 The temperature demand in Cambridge, 2006, visualized as an area on a graph of daily average temperatures. (a) Thermostat set to 20℃, including cooling in summer; (b) winter thermostat set to 17℃.

$$\text{energy lost} = \text{leakiness} \times \text{temperature demand}.$$

We can reduce our energy loss by reducing the leakiness of the building, or by reducing our temperature demand, or both. The next two sections look more closely at these two factors, using a house in Cambridge as a case-study.

There is a third factor we must also discuss. The lost energy is replenished by the building's heating system, and by other sources of energy such as the occupants, their gadgets, their cookers, and the sun. Focussing on the heating system, the energy delivered by the heating is not the same as the energy consumed by the heating. They are related by the coefficient of performance of the heating system.

For a condensing boiler burning natural gas, for example, the coefficient of performance is 90%, because 10% of the energy is lost up the chimney.

To summarise, we can reduce the energy consumption of a building in three ways:
1. by reducing temperature demand;
2. by reducing leakiness; or
3. by increasing the coefficient of performance.

Resource for reference

David JC MacKay. *Sustainable Energy-without the Hot Air*. 2009: UIT Cambridge Ltd.

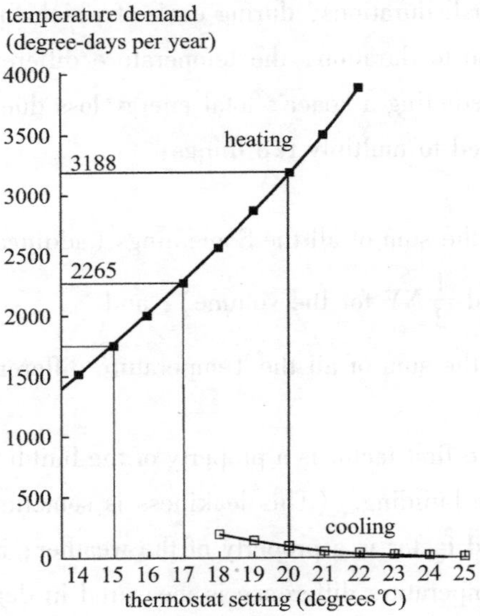

Figure 33-24 Temperature demand in Cambridge, in degree-days per year, as a function of thermostat setting (℃). Reducing the winter thermostat from 20℃ to 17℃ reduces the temperature demand of heating by 30%, from 3188 to 2265 degree-days. Raising the summer thermostat from 20℃ to 23℃ reduces the temperature demand of cooling by 82%, from 91 to 16 degree-days.

Section 3
Tips for Translation

汉译英中名词修饰语的排列

把有两个或两个以上的修饰语的名词词组翻译成英语的时候，必须注意英语中修饰语的排列习惯。

1. 意大利著名的建筑师
a famous Italian architect

2. 美国的知名建筑院校
a renowned American architecture school

3. 中国杰出的建筑史学家
an outstanding Chinese historian of architecture

　　一般而言，在英语中名词性的修饰语往往置于形容词性的修饰语之后，而在汉语中刚好相反。上面三个例子就是如此。

4. 大型结构构件
a large structural component

5. 统一的创造过程
a unified creative process

6. 漂亮的曲面形式
a beautiful curved shape

　　如果修饰语均为描绘性的形容词，首先应当根据原文的逻辑关系进行翻译，如上面的例子。但是如果这些形容词是完全并列的关系时，在翻译时还必须注意英语的一个语言习惯，即根据词的长短以及第一个字母在字母表中的排列顺序确定先后关系。在这种情况下，英语翻译同汉语原文就不必再位置上一一对应了。如"勤劳勇敢的中国人民"可译为"the brave, hard-working Chinese people"，"贫穷落后的国家"译为"a poor, under-developed country"，等等。

　　在汉语名词修饰语的翻译中还有一种情况，即用后置的同位语修饰语来翻译在汉语中前置的修饰语。如：

7. 普里茨克建筑奖获得者菲利浦·约翰逊同他的合伙人约翰·伯吉一起设计了休斯敦的美国银行大楼，这是一栋采用粉色花岗石、顶部为退台式坡屋顶的56层塔楼。
Philip Johnson, a Pulitzer Architecture Prize winner, with his partner, John Burgee, designed the Bank of America building in Houston, a 56-story tower of pink granite stepped back in a series of Dutch gable roofs.

8. 著名的建筑师、结构工程师和雕塑家圣地亚哥·卡拉特拉瓦以其独特的设计美学赢得了世界范围的声誉。
Santiago Calatrava, a world-renowned architect, engineer and sculptor, gained his fame around the world for his unique design aesthetic.

Unit 34

Section 1
Intensive Reading

"Im Birch" School

Part II

Design I

The expression of the facade is characterised by the precast concrete lesenes which divide up the sur-faces vertically. These elements are not merely decorative but since they are also employed for fixing the windows. All facades use this system——the classroom wings and the sports hall.

The use of different precast concrete elements for items such as the roof edge, lesenes, slab edge and plinth leads to a calm, static, almost classical facade construction. The spacing of the lesenes is equal to half the distance between the grid lines of the structural layout, which permits the use of different materials for the in-fill panels: glass, rendered surfaces, other concrete elements or steel features (safety barriers along the escape balconies). These disparate infills alter the references to the surrounding spaces (Figure 34-1 ~ Figure 34-5).

Figure 34-1 Lesenes Precast concrete element Lesene (pilaster strip) in this sense is a pier- type projecting strip of wall without a capital or a base.

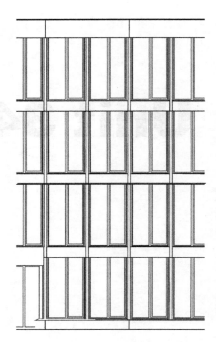

Figure 34-2 Elevation, 1:50

Figure 34-3 Corner of building and edge of roof Steel safety barrier to escape balcony, glazing and rendered brick wall

Roof construction
Rooftop planting (unplanned, i. e. natural)
substrate, optional drainage mat   100mm
Separating layer
Waterproofing, root-resistant EP 4   10mm
Mineral-fibre insulati   200mm
Vapour barrier, VA 4, fully bonded   10mm
Fair-face concrete slab laid to falls   260mm
Acoustic ceiling panel   70mm
Total   650mm

Floor construction, upper floors
Linoleum/adhesive   5mm
Optionalvapour check
Fibre-reinforced screed   25mm
Fair-face concrete slab   340mm
Acoustic ceiling panel (perforated and   70mm
painted gypsum boards)
Luminaires
Total   440mm

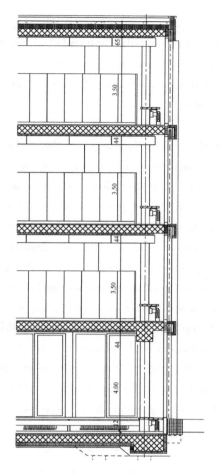

Figure 34-4 Section, 1:50

Figure 34-5 Building housing primary school and sports hall

Floor construction, ground floor	
Linoleum/adhesive	5mm
Fibre-reinforced screed	85mm
or	
(12mm stone flags laid in adhesive	
78mm screed on separating layer)	
Mineral-fibre impact sound insulation	20mm
Thermal insulation, expanded polystyrene (F20)	40mm
Vapour check	10 mm
Concrete slab, waterproof	300mm
Polyethylene sheet,	0.2mm
Thermal insulation, extruded polystyrene	120mm
Blinding layer, lean concrete	
Total	580mm

Unit 34 | 259

Design II

In comparison with existing structures that employ the column-and-slab principle (e. g. Le Corbusier's Dom-Ino principle), the architect exploits neither the independent arrangement of the facade, nor the freedom in the internal layout that would be possible. Instead, this system can be regarded as a neutral framework for the structure of the facade.

The clear and simple assembly of the individual concrete elements is dominated by the prefabrication and the logistics of the erection. The first phase involves insulating the edges of the floor slabs and attaching angles ready for fixing the windows later. At ground floor the edges of the slabs include nibs measuring 440mm × 330mm × 300mm on which the prefabricated plinth elements are seated. The lesenes are fixed, storey by storey, to the load bearing structure, i. e. to the edges of the concrete slabs. The horizontal concrete elements——to conceal and protect the sunblinds and form sills for the windows above——are then mounted on the lesenes. The roof edge elements are fixed with Omega expansion anchors, while loadbearing facade anchors with spacer bolts are used for the lesenes. The brick infill panels are built up in situ. In the second phase the thermal insulation is attached. This consists of storey-high elements of 220 mm thick expanded polystyrene which are bonded directly to the brickwork and subsequently rendered. The aluminium windows are mounted between the lesenes on the angle sections that were attached earlier. All precast concrete elements have open, drained joints, i. e. the design of the individual elements and the logic of their jointing obviates the need for sealing materials (Figure 34 – 6 ~ Figure 34 – 10).

The storey-high openings within the grid of lesenes are divided in two. Each window consists of a fixed light and a bottom-hung light. A controlled air-conditioning system with heat recovery has been installed and complies with Switzerland's "Minergie Standard". The classrooms are fitted

Figure 34-6 View of slab soffit showing lesene with facade anchor and spacer bolt supporting vertical concrete element

Figure 34-7 View during construction Lesenes and vertical concrete elements, masonry infill panels

Figure 34-8   Plinth Cantilevering ground floor slab and plinth zone with nibs

Figure 34-9   Facade to secondary school Window with fixed light and bottom-hung light, fixed desks and radiators

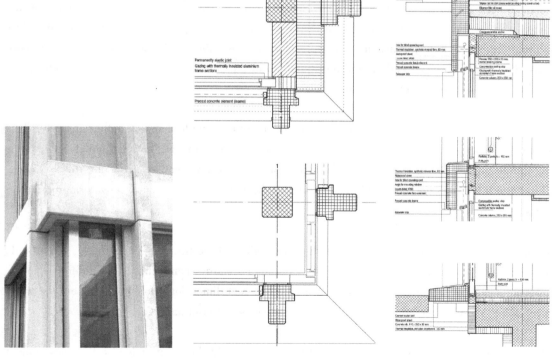

Figure 34-10   Close-up of facade Corner between ground and 1st floor levels on primary school building

with built-in cupboards for the necessary teaching materials; the cupboards have fresh-air inlets at the base. Exhaust air is extracted via a duct that runs above the suspended ceiling along the inside wall.

All classrooms are fitted with internal blackout blinds or curtains that run in tracks along the glazed system walls to the common areas. The classrooms can also be darkened by means of louvre blinds. Another feature is the built-in tables fixed between the columns which also serve to conceal the radiators. Services run in the duct along the spandrel panel below the windows.

Materials
The precast concrete elements (lesenes, spandrel panels and plinth segments), the grey render and the anodised aluminium windows form the visible elements of the building on the outside. These materials and the way they are used essentially determine the colour scheme, or rather the restrained "colouring" of the complex, with the areas of glass, which appear dark, plus the dark render contrasting with the light colouring of the concrete elements in the facade.

Inside, the loadbearing structure of the building is always present. The uneven, raw texture of the fair-face concrete surfaces is finished with a clear lacquer, which gives the walls a stony appearance. Non-loadbearing parts complement the structural elements: the glass block walls, the glazing and the brick walls, finished with white-painted glass-fibre wallpaper. Whereas open-pore travertine flags have been laid around the stairs and in the entrance lobbies, beige-coloured linoleum has been used in the teaching units.

The materials employed and their different surface qualities seem to converge rather abruptly. This suggests a pragmatic approach: established rules, whether in terms of jointing the materials, framing the glazing or detailing the plinth, are part of an overriding plan of action by the architect. They form a tool for the controlled management of the planning work, an approach appropriate to the size of the building.

To take as an example the edge detail for the stair flight and the travertine stair finish, the attitude of the architects with respect to jointing the materials is readily seen. The actual difference in the accuracy of the materials is allowed for, i.e. the different dimensional tolerances govern the treatment of the in situ concrete, which becomes an obvious, protruding edge.

The architectural allocation and presence of the elements and their materials also becomes evident in the routing of the building services. The horizontal distribution along the floor slabs takes place in a duct with branches, an efficient method, and around the lobbies along the edges of the floor slabs. The services duct is clad with grey sheet metal and appears to be trying to find its way along the floors in order to supply all the classrooms (Figure 34-11 ~ Figure 34-13).

Figure 34-11 Close-up of stairs Joint between in situ concrete and travertine

Figure 34-12 Entrance area for dining and sports halls The columns in the entrance areas are clad with travertine.

Words and phrases

1. nibs   *n.*   大人物；上司；傲慢的人
2. sills   *n.*   ［建］基石（sill 的复数）
3. anchor *n.*   锚
4. spacer *n.*   垫片
5. exhaust air   废气；排气
6. masonry *n.*   石工；石工行业；石造建筑
7. parapet *n.*   栏杆；扶手；矮护墙
8. spout   *n.*   喷口；水龙卷；水落管；水柱
9. bitumen *n.*   沥青
10. fascia 筋膜
11. recess *n.*   凹处   *vt.*   使凹进
12. putty *n.*   油灰；氧化锡   *vt.*   用油灰填塞
13. spandrel *n.* 拱肩
14. lacquer *n.*   漆；漆器   *vt.*   涂漆；使表面光泽

Figure 34-13 Materials in classroom Fair-face concrete soffit and acoustic gypsum panels with integral lighting units, ventilation duct clad with grey sheet metal, glazing with curtains, peripheral aluminium rail for displaying drawings etc., sink and blackboard, beige-coloured linoleum floor covering

Section 2
Extensive Reading

Heating

Part II

Temperature demand

We can visualize the temperature demand nicely on a graph of external temperature versus time (Figure 34-14). For a building held at a temperature of 20℃, the total temperature demand is the area between the horizontal line at 20℃ and the external temperature. In Figure 34-14, we see that, for one year in Cambridge, holding the temperature at 20℃ year-round had a temperature demand of 3188 degree-days of heating and 91 degree-days of cooling. These pictures allow us easily to assess the effect of turning down the thermostat and living without air-conditioning. Turning the winter thermostat down to 17℃, the temperature demand for heating drops from 3188 degree-days to 2265 degree-days (Figure 34-14), which corresponds to a 30% reduction in heating demand. Turning the thermostat down to 15℃ reduces the temperature demand from 3188 to 1748 degree days, a 45% reduction.

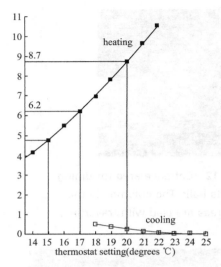

Figure 34-14 The temperature demand in Cambridge, 2006, replotted in units of degree-days per day, also known as degrees. In these units, the temperature demand is just the average of the temperature difference between in side and outside.

These calculations give us a ballpark indication of the benefit of turning down thermostats, but will give an exact prediction only if we take into account two details: first, buildings naturally absorb energy from the sun, boosting the inside above the outside temperature, even without any heating; and second, the occupants and their gadget companions emit heat, so further cutting down the artificial heating requirements. The temperature demand of a location, as conventionally expressed in degree-days, is a bit of an unwieldy thing. I find it hard to remember numbers like "3500 degree-days". And academics may find the degree-day a distressing unit, since they already have another meaning for degree days (one involving dressing up in gowns and mortar boards). We can make this quantity more meaningful and perhaps easier to work with by dividing it by 365, the number of days in the year, obtaining the temperature demand in "degree-days per day", or, if you prefer, in plain "degrees". Figure E.6 shows this replotted temperature demand. Expressed this way, the temperature demand is simply the average temperature difference between inside and outside. The highlighted temperature demands are: 8.7℃, for a thermostat setting of 20℃; 6.2℃, for a setting of 17℃; and 4.8℃, for a setting of 15℃.

Leakiness-example: my house

My house is a three-bedroom semi-detached house built about 1940 (Figure 34-15). By 2006, its kitchen had been slightly extended, and most of the windows were double-glazed. The front door and back door were both still single-glazed.

Figure 34-15 My house.

My estimate of the leakiness in 2006 is built up as shown in table 34-1. The total leakiness of the house was 322W/℃ (or 7.7kWh/d/℃), with conductive leakiness accounting for 72% and ventilation leakiness for 28% of the total. The conductive leakiness is roughly equally divided into three parts: windows; walls; and floor and ceiling.

Conductive leakiness	area (m^2)	U-value (W/m^2/℃)	leakiness (W/℃)
Horizontal surfaces			
Pitched roof	48	0.6	28.8
Flat roof	1.6	3	4.8
Floor	50	0.8	40
Vertical surfaces			
Extension walls	24.1	0.6	14.5
Main walls	50	1	50
Thin wall (5in)	2	3	6
Single-glazed doors and windows	7.35	5	36.7
Double-glazed windows	17.8	2.9	51.6
Total conductive leakiness			232.4

Table 34-1. Breakdown of my house's conductive leakiness, and its ventilation leakiness, pre-2006. I've treated the central wall of the semi-detached house as a perfect insulating wall, but this may be wrong if the gap between the adjacent houses is actually well-ventilated. I've highlighted the parameters that I altered after 2006, in modifications to be described shortly.

Ventilation leakiness	volume(m^3)	N(air-changes per hour)	leakiness(W/℃)
Bedrooms	80	0.5	13.3
Kitchen	36	2	24
Hall	27	3	27
Other rooms	77	1	25.7
Total ventilation leakiness			90

To compare the leakinesses of two buildings that have different floor areas, we can divide the leakiness by the floor area; this gives the heat-loss parameter of the building, which is measured in W/℃/m^2. The heat-loss parameter of this house (total floor area 88m^2) is 3.7W/℃/m^2. Let's use these figures to estimate the house's daily energy consumption on a cold winter's day, and year-round. On a cold day, assuming an external temperature of 1℃ and an internal temperature of 19℃, the temperature difference is $\Delta T = 20$℃. If this difference is maintained for 6 hours per day then the energy lost per day is

$$322\text{W/℃} \times 120 \text{ degree-hours} \cong 39\text{kWh}.$$

If the temperature is maintained at 19℃ for 24 hours per day, the energy lost per day is 155kWh/d.

To get a year-round heat-loss figure, we can take the temperature demand of Cambridge from figure E.5. With the thermostat at 19℃, the temperature demand in 2006 was 2866 degree-days. The average rate of heat loss, if the house is always held at 19℃, is therefore:

$$7.7\text{kWh/d/℃} \times 2866 \text{ degree-days/y}/(365 \text{ days/y}) = 61\text{kWh/d}.$$

Turning the thermostat down to 17℃, the average rate of heat loss drops to 48kWh/d. Turning it up to a tropical 21℃, the average rate of heat loss is 75kWh/d.

Effects of extra insulation

During 2007, I made the following modifications to the house:
1. Added cavity-wall insulation (which was missing in the main walls of the house).
2. Increased the insulation in the roof.
3. Added a new front door outside the old.
4. Replaced the back door with a double-glazed one.
5. Double-glazed the one window that was still single-glazed. What's the predicted change in heat loss?

The total leakiness before the changes was 322W/℃.

Adding cavity-wall insulation (new U-value 0.6) to the main walls reduces the house's leakiness by 20W/℃. The improved loft insulation (new U-value 0.3) should reduce the leakiness by 14W/℃. The glazing modifications (new U-value 1.6–1.8) should reduce the conductive leakiness by 23W/℃, and the ventilation leakiness by something like 24W/℃. That's a total reduction in leakiness of 25%, from roughly 320 to 240W/℃ (7.7 to 6kWh/d/℃). Table E.9 shows the predicted savings from each of the modifications.

The heat-loss parameter of this house (total floor area 88m^2) is thus hopefully reduced by about 25%, from 3.7 to 2.7W/℃/m^2. (This is a long way from the 1.1W/℃/m^2 required of a "sustainable" house in the new building codes.)

Break-down of the predicted reductions in heat loss from my house, on a cdd winter day.   Table 34–2

-Cavity-wall insulation (applicable to two-thirds of the wall area)	4.8kWh/d
-Improved roof insulation	3.5kWh/d
-Reduction in conduction from double-glazing two doors and one window	1.9kWh/d
-Ventilation reductions in hall and kitchen from improvements to doors and windows	2.9kWh/d

It's frustratingly hard to make a really big dent in the leakiness of an already-built house! As we saw a moment ago, a much easier way of achieving a big dent in heat loss is to turn the thermostat down. Turning down from 20 to 17℃ gave a reduction in heat loss of 30%.

Combining these two actions——the physical modifications and the turning-down of the thermostat——this model predicts that heat loss should be reduced by nearly 50%. Since some heat is generated in a house by sunshine, gadgets, and humans, the reduction in gas consumption should be more than 50%.

I made all these changes to my house and monitored my meters every week. I can confirm that my heating bill indeed went down by more than 50%.

Resource for reference

David JC MacKay. *Sustainable Energy-without the Hot Air*. 2009: UIT Cambridge Ltd.

Section 3
Tips for Translation

汉译英中的顺译法、变序法和并句法

  顺序法、变序法和并句法也是长句翻译时候常见的方法。

1. 索拉里曾被授予三次荣誉博士学位，1963 年美国建筑师学会金奖，1981 年保加利亚索非亚世界建筑双年展金奖，以及 1984 年法国建筑学院银奖。
Soleri has been awarded three honorary doctorates, the American Institute of Architects Gold Medal in 1963, the Gold Medal from the World Biennieal of Architecture in Sofia, Bulgaria, in 1981, and the Silver Medal of the Academied´Architecture in France, 1984.

2. 板的刚度就像梁一样取决于其厚度，如果太薄，板就会由于太容易变形而失去作用。
The stiffness of flat slabs, like that of beams, derives from their thickness: if too thin they become too flexible to be functional.

  1、2 两句基本按照汉语的句子顺序和逻辑顺序翻译，即"顺序法"翻译。

3. 世界范围内住宅设计正越来越引起建筑师们关注，因此我们相信对 21 世纪一些伟大的住宅项目的重新审视是有必要的。
It is believed that a reexamination of some of the great housing projects of this century is appropriate at a time when the design of housing is drawing more and more attention from architects around the world.

4. 如果没有伴随结构方式的创新，我最有趣的两个项目：采用预制构件的飞机库和都灵展览馆的屋顶，是不可能实现的。
Two of my most interesting projects, the hangars built of pre-cast elements and the roof for the Turin Exhibition Halls, would have been impossible without a simultaneous invention of the structural method.

3、4两句的翻译在力求表述汉语原文本意的前提下,结合英语的表达习惯和思维方式,对汉语原文句子成分的前后顺序进行适当调整,使翻译顺畅、紧凑,即采取"变序法"。

5. 他用在一家建筑设计公司做五年的绘图员存的钱买了一架照相机。然后带着它走遍欧洲,参观了大量的著名建筑,拍摄了无数的照片,这对他今后的建筑师生涯十分重要。
With the money he saved for five years by working as an draftsman in an architecture design company, he bought a camera, with which he traveled around Europe, visiting a lot of famous architecture and taking numerous photos, which was proved to be very important by his later career as an architect.

6. 要明白建筑是如何造出来的,我们必须了解建筑构造的一些知识。正是凭借建筑构造,不同的结构构件和建筑材料被组合在一起,创造出生活的空间。
To understand how architectures is built, we must first have theknowledge of building construct, through which different structural elements and building materials are put together with each other to create living space.

汉语中经常会出现一些并置的短句,虽然在语法结构上彼此独立,但意义上关系密切。在汉译英的时候,可以采用各种从句、短语等连接手段,翻译成一句话,即采取"并句法"。5、6两句的翻译就是如此,汉译英时的"并句法"同英译汉时的对英语长句的分断处理(参见"Translating long sentences 长句的翻译")实际上是一种对应关系。

Unit 35

Section 1
Intensive Reading

DETACHED FAMILY HOME

Part I

Situation and theme

Grabs, the kind of scattered settlement, that is typical in Switzerland, lies in the flat land of the St Galler Rhine valley. Peter Markli's house stands in a gentle depression between farms and other detached houses. It faces south and access is from the north side, via a narrow asphalt road (Figure 35–1 ~ Figure 35–3).

Figure 35–1 Site plan

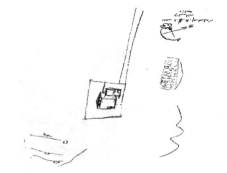

Figure 35–2 Sketch showing location and context

At the start the design work was marked by an intensive analysis of the location and the interior layout, always keeping in mind the needs of the occupants. In the course of the design process the

Figure 35-3 The house stands like sculpture on the open ground.

aim was to focus on a few themes——"one decides in favour of a whole". One sketch finally embodied all the essential factors of the design.

Markli responded to the given situation with a solitary, compact building. The house does not attempt to fit in with the existing buildings; it distances itself, so to speak, from its environment. It achieves this through abstraction. The intent here is not "minimal art" or a "new simpleness", but rather a directness of expression in which all parts of the whole are visualised together.

Relationship with the terrain

The open ground on which the house is built had to remain intact as far as possible. Therefore, the cantilevering part of the veranda seems to float above the ground. All the elements grow out of the envelope itself, which lends the building an autonomous, even introverted expression. It was not intended to be a house with external facilities competing with the neighbouring farmyards. The house is different from its surroundings, or as Ines Lamuniere says: "It possesses a certain austerity which confines people either to the inside or the outside." A private garden in the normal sense of the word would be inconceivable here; the private external space——the veranda——is part of the house (Figure 35-4 ~ Figure 35-6).

Figure 35-4 The veranda is seemingly cut out of the veloum.

Figure 35-5 The veranda "floats" above the ground.

Figure 35-6 The veranda-external and yet enclosed

Interior layout

The plan evolved around a focal point along the lines of the "onion skin principle". A few steps lead up from the covered entrance area to the hall, from where stairs lead to the upper floor and basement. The living room and kitchen are arranged in an L-shape on two sides of the hall. The large sliding windows allow a good view of the veranda and the seemingly distant surroundings beyond. The sliding aluminium shutters, providing privacy and protection from direct sunlight, help to reinforce this effect. Owing to the relationship between the corner and a section of wall, the interior space becomes opened up. This space then, devoid of any intervening columns, with the folding dividing wall between kitchen and living room, and a cement screed floor finish throughout, achieves an astounding expansiveness.

The interior layout on the upper floor also makes use of the L-shape. The south-facing rooms in the "L" are reached from a central hall, brightly lit via rooflights. The rooms, cantilevering out over the veranda, are of different sizes and are separated by plasterboard walls and built-in cupboards. The tiled bathrooms have been placed on the north side of the building (Figure 35 – 7 ~ Figure 35 – 10).

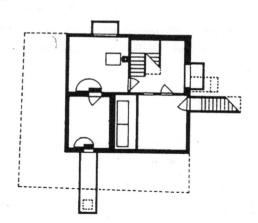

Figure 35–7   Plan of upper floor

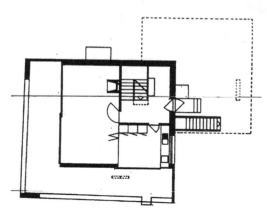

Figure 35–8   Plan of ground floor

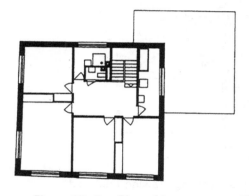

Figure 35–9   Plan of basement

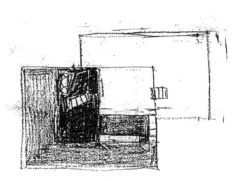

Figure 35–10   Sketch showing interlacing of rooms

Construction and structural aspects

The use of in situ concrete is underscored by the nonright-angled geometry of the building, "which allows the cast form to be seen as bordering on the ideal, so to speak". The homogeneity of the cube is achieved by a constructional separation. The outer skin of concrete is structurally independent, with the loads being carried through prestressing and cantilevers. The inner skin is of plastered masonry. The concrete wall at ground floor level is the sole free-standing structural element. Besides its loadbearing function, it lends structure to the plan layout and marks the limit of the living room.

The inner skin, masonry and concrete floors could be removed at a later date; the outer concrete envelope is totally separate from these in a structural sense. The point in the floor slab over the ground floor where the inner and outer skins meet (circled in 3) is the point at which the large sliding windows to the veranda are incorporated. The use of such large window elements, without employing any cover strips, required a high degree of precision (tight tolerances) during manufacture and installation (Figure 35 – 11 ~ Figure 35 – 15).

Figure 35–11   Entrance elevation

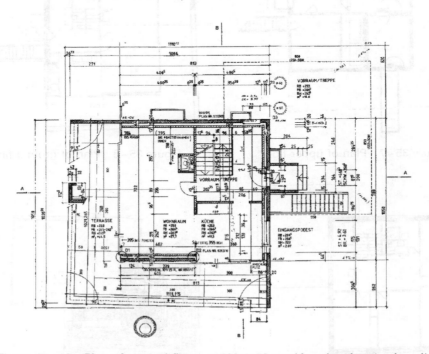

Figure 35–12   Plan of ground floor, 1∶100 1∶50 working drawing (reduced)

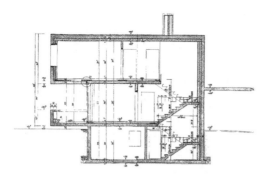
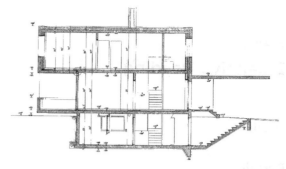

Figure 35–13 Section A-A, 1:100 1:50 working drawing (reduced)

Figure 35–14 Section B-B, 1:100

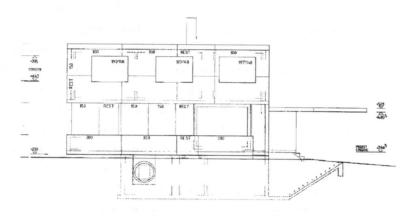

Figure 35–15 South elevation, 1:100

Words and phrases

1. detached   *adj.* 分离的，分开的；超然的   *v.*   分离
2. asphalt   *n.*   沥青；柏油   *vt.*   以沥青铺   *adj.*   用柏油铺成的
3. solitary   *n.*   独居者；隐士   *adj.* 孤独的；独居的
4. terrain   *n.*   ［地理］地形，地势；领域；地带
5. intact   *adj.*   未触动的，未碰过的；未变质的
6. veranda   *adj.*   游廊，走廊，凉台间
7. autonomous   *adj.*   自治的 独立的
8. introverted   *adj.*   内向的   *v.*   使…内向（introvert 的过去分词形式）；使内翻
9. austerity   *n.*   紧缩；朴素；苦行；严厉
10. focal   *adj.*   焦点的，在焦点上的；灶的，病灶的
11. basement   *n.*   地下室；地窖
12. intervening   *adj.*   介于中间的
13. astounding   *adj.*   令人震惊的；令人惊骇的
14. homogeneity   *n.*   同质；同种；同次性

Section 2
Extensive Reading

Heating

Part III

Air-exchange

Once a building is really well insulated, the principal loss of heat will be through ventilation (air changes) rather than through conduction. The heat loss through ventilation can be reduced by transferring the heat from the outgoing air to the incoming air. Remarkably, a great deal of this heat can indeed be transferred without any additional energy being required. The trick is to use a nose, as discovered by natural selection. A nose warms incoming air by cooling down outgoing air. There's a temperature gradient along the nose; the walls of a nose are coldest near the nostrils. The longer your nose, the better it works as a counter-current heat exchanger. In nature's noses, the direction of the air-flow usually alternates. Another way to organize a nose is to have two air-passages, one for in-flow and one for out-flow, separate from the point of view of air, but tightly coupled with each other so that heat can easily flow between the two passages. This is how the noses work in buildings. It's conventional to call these noses heat-exchangers.

Figure 35-16 The Heatkeeper Serrekunda.

An energy-efficient house

In 1984, an energy consultant, Alan Foster, built an energy-efficient house near Cambridge; he kindly gave me his thorough measurements. The house is a timber-framed bungalow based on a Scandinavian "Heatkeeper Serrekunda" design (Figure 35-16), with a floor area of $140m^2$, composed of three bedrooms, a study, two bathrooms, a living room, a kitchen, and a lobby. The wooden outside walls were supplied in kit form by a Scottish company, and the main parts of the house took only a few days to build.

The walls are 30 cm thick and have a U-value of $0.28 W/m^2/°C$. From the inside out, they consist of 13mm of plasterboard, 27mm airspace, a vapour barrier, 8mm of plywood, 90mm of rockwool, 12mm of bitumen-impregnated fibreboard, 50mm cavity, and 103mm of brick. The ceiling construction is similar with 100~200mm of rockwool insulation. The ceiling has a U-value of $0.27 W/m^2/°C$, and the floor, $0.22 W/m^2/°C$. The windows are double-glazed (U-value $2W/m^2/°C$), with the inner panes' outer surfaces specially coated to reduce radiation. The windows are arranged to give substantial solar gain, contributing about 30% of the house's space-heating.

The house is well sealed, every door and window lined with neoprene gaskets. The house is heated by warm air pumped through floor grilles; in winter, pumps remove used air from several rooms, exhausting it to the outside, and they take in air from the loft space. The incoming air and outgoing air pass through a heat exchanger (Figure 35–17), which saves 60% of the heat in the extracted air. The heat exchanger is a passive device, using no energy: it's like a big metal nose, warming the incoming air with the outgoing air. On a cold winter's day, the outside air temperature was $-8°C$, the temperature in the loft's air intake was $0°C$, and the air coming out of the heat exchanger was at $+8°C$.

For the first decade, the heat was supplied entirely by electric heaters, heating a 150-gallon heat store during the overnight economy period. More recently a gas supply was brought to the house, and the space heating is now obtained from a condensing boiler.

Figure 35–17 The Heatkeeper's heat-exchanger.

The heat loss through conduction and ventilation is $4.2 kWh/d/°C$. The heat loss parameter (the leakiness per square metre of floor area) is $1.25 W/m^2/°C$ (cf. my house's $2.7 W/°C/m^2$).

With the house occupied by two people, the average space-heating consumption, with the thermostat set at 19°C or 20°C during the day, was 8100kWh per year, or 22kWh/d; the total energy consumption for all purposes was about 15000kWh per year, or 40kWh/d. Expressed as an average power per unit area, that's $6.6 W/m^2$.

Figure E.12 compares the power consumption per unit area of this Heatkeeper house with my house (before and after my efficiency push) and with the European average. My house's post-efficiency-push consumption is close to that of the Heatkeeper, thanks to the adoption of lower thermostat settings.

Resource for reference

David JC MacKay. *Sustainable Energy-without the Hot Air*. 2009：UIT Cambridge Ltd.

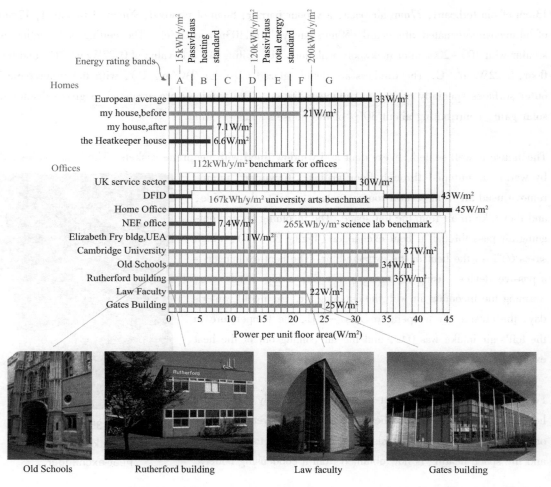

Figure 35-18 Building benchmarks. Power used per unit area in various homes and offices.

Section 3
Tips for Translation

汉译英中时主动句译为被动句

  汉译英时的主动句译为被动句同英译汉时的被动句译为主动句（参见"Translating the passive voice 被动语态的翻译"）实际上也是一种对应关系. 而且在很多时候汉语并不一定

要用"被"、"受"、"让"等助词来表达被动含义。在这种情况下,虽然汉语原文本身不是被动句式,但在英译过程中采用被动句式把汉语的被动含义体现出来。

1. 世界最高的摩天楼将在年底前建成。
The tallest skyscraper in the world will be completed by the end of this year.

2. 华莱士·纳夫最先在1940年代使用充气结构。
Pneumatic forms were first used in the 1940s by Wallace Neff.

3. 如今越来越多的人认识到:人工城市中缺乏某些根本的要素。
It is more and more widely recognized today that there is some essential ingredient missing from artificial cities.

4. 事故造成17人死亡. 于是公众开始呼吁铁路系统的电气化。
17 People were killed in the accident and a public outcry called for electrification of the railway system.

5. 把汽车引入车站周边区域的架空坡道使人车分离。
Pedestrians and cars are separated by special elevated ramps which lead the cars around the railway station.

6. 天花板画满了法国艺术家鲍尔·赫勒创作的作品. 十二星象的构思来取自中世纪的手稿。
The ceiling is painted by the French artist Paul Helleu. The design with zodiac constellations was taken from a medieval manuscript.

　　汉译英时的主动句常常被译为被动句,同时也是因为英语中被动句较多而汉语中被动句较少。特别是在学术文献中,为了体现客观性,被动句式更是常见。

Section 4
Tips for Writing

Use the active form
Active verb forms can grab and retain the reader's interest. "Jack loves Jane" sparkles; the passive form "jane is loved by jack" is dull. The same holds true when design professionals and students write. Apply this consistently, and you'll greatly sharpen your impact, as you can see in these before and after examples.

It is intended that AutoCAD 2005, Form Z-4.0, and 3D StudioMax will be utilized and files plotted to devices using Windows NT Print Manager drivers. (*from a proposal*) Try turning this sentence around, then check your result with the new version.

We intend to use AutoCAD 2005, Form Z-4.0, and 3D StudioMax and to plot files to devices using Window NT Print Manager drivers.

5. Don't forget people

If you intend to capture the interest of a selection committee or a powerful patron, you need to mix in with your writing a generous measure of human references. Clients are more likely to identify with a message if it is styled with people involved, rather than as a dry-as-dust task by an anonymous presence. Avoid this example from a proposal.

Past costs and schedules on comparative projects are analyzed in order to achieve viable final cost estimates and realistic completion dates.

Try the following instead.

Jane Smith and her staff analyze past projects to arrive at realistic final cost estimates for your project. Our present management support staff, lead by George Lopez, review previous job schedules to come up with a feasible schedule for our clients.

Resource for reference

Stephen A. Kliment, 1998, Writing for Design Professionals, W. W. Norton & Company, New York, USA.

Unit 36

Section 1
Intensive Reading

DETACHED FAMILY HOME

Part II

Facades

Here again there is no clear hierarchy among the components. As with the interior layout the most important thing in this case is the proportions. The relationship between the parts and the whole, between the parts themselves, and between openings and wall surfaces are crucial influences on the expression of the building. Internally, Markli also controls the elevations and the positions of openings in every single room by means of a consistent system of dimensions. At the lowest hierarchic level we have the pattern of formwork joints, which itself is subservient to the surface (Figure 36 – 1 ~ Figure 36 – 4).

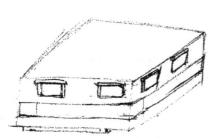

Figure 36 – 1   Sketch showing facade proportions

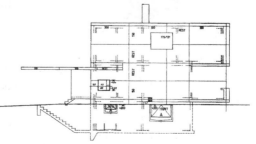

Figure 36 – 2   East elevation, 1:100

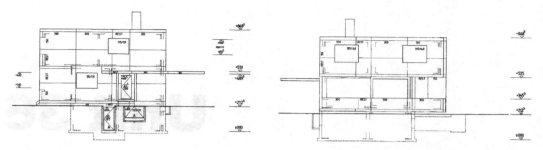

Figure 36-3　North elevation, 1:100　　　　Figure 36-4　West elevation, 1:100

Small sketches showing two elevations were used to check the relationships.

Markli works according to visual rules. The north elevation, for instance, is dominated by the two divergent cantilevers——the canopy over the entrance area and the veranda——and these add a certain tension to the facade. But the openings are positioned in such a way that the visual balance is restored. What this means is that the "centre of gravity" for the viewer comes to rest within the outline of the building (one can check this with the view towards the corner).

A single element like the long cantilevering canopy always has more than one function. Besides the architectural use already mentioned, it also serves as a symbol for the entrance, protects the entrance from the weather and acts as a carport.

Openings

For tectonic reasons, the windows finish flush with the outside face, which helps to emphasise the coherence of the envelope. This results in deep internal reveals, whose "archaic" nature would not normally suit the character of such a house. Markli solves this problem by including a wooden lining on the inside with a recess for storing the shutters. With the lighting units also being positioned above the window, the technical elements are concentrated around the opening. The walls and ceilings therefore remain intact, a coherent whole (Figure 36-5 ~ Figure 36-7).

Figure 36-5　Window flush with facade surface Fitting the window in this way calls for carefully controlled details in terms of sealing against driving rain and wind pressure (rebated joints).

There are two different types of window, in both cases horizontal pivot windows in aluminium frames. In the rooms above the cantilevering veranda the "wooden box", fitted with folding shutters of imitation leather, projects into the room. On the north side, in the kitchen and in the bathrooms, this box is fitted flush with the inside wall. It houses painted folding wooden shutters to provide privacy and protection against direct sunlight. All the folding shutters are standard products easily integrated into the whole thanks to their accurate design and fabrication (Figure 36-8 ~ Figure 36-17).

Figure 36-6 Aluminium horizontal pivot window

Figure 36-7 Window with "imitation leather bellows"

Figure 36-8 Horizontal pivot window from outside

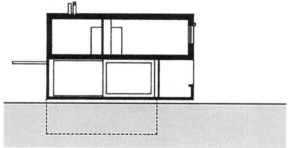

Figure 36-9 Window type II

Figure 36-10 "Imitation leather bellows" from inside

Figure 36-11 "Imitation leather bellows" from outside

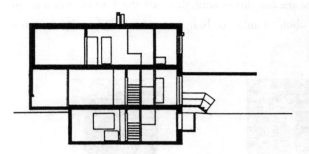

Figure 36-12　Window type I

Figure 36-13　Horizontal pivot window from inside

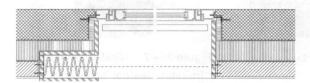

Figure 36-14　Horizontal section, 1:10

Figure 36-15　Horizontal section, 1:10

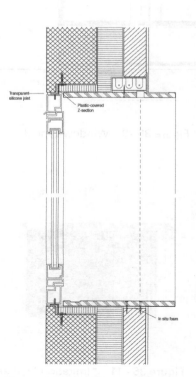

Figure 36-16　Section through window, 1:10

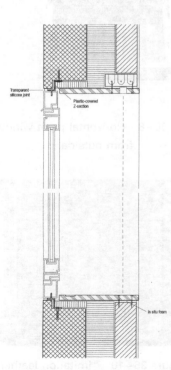

Figure 36-17　Section through window, 1:10

Words and phrases

1. hierarchy *n.* 层级；等级制度
2. proportion *n.* 比例；大小（proportion 的复数形式）
3. subservient *adj.* 屈从的；奉承的；有用的；有帮助的
4. divergent *adj.* 相异的，分歧的；散开的
5. canopy *n.* 天篷；华盖；遮篷；苍穹 *vt.* 用天篷遮盖；遮盖
6. carport *n.* 车库；（美）车棚
7. tectonic *adj.* 构造的；建筑的；地壳构造上的
8. archaic *adj.* 古代的；陈旧的；古体的；古色古香的
9. pivot *n.* 枢轴；
10. shutter *n.* 百叶门窗，百叶窗

Section 2
Extensive Reading

Heating

Part IV

Benchmarks for houses and offices

The German Passivhaus standard aims for power consumption for heating and cooling of $15kWh/m^2/y$, which is $1.7W/m^2$; and total power consumption of $120kWh/m^2/y$, which is $13.7W/m^2$. The average energy consumption of the UK service sector, per unit floor area, is $30W/m^2$.

An energy-efficient office

The National Energy Foundation built themselves a low-cost low-energy building. It has solar panels for hot water, solar photovoltaic (PV) panels generating up to 6.5kW of electricity, and is heated by a 14kW ground-source heat pump and occasionally by a wood stove. The floor area is $400m^2$ and the number of occupants is about 30. It is a single-storey building. The walls contain 300mm of rockwool insulation. The heat pump's coefficient of performance in winter was 2.5. The energy used is 65kWh per year per square metre of floor area ($7.4W/m^2$). The PV system delivers almost 20% of this energy.

Contemporary offices

New office buildings are often hyped up as being amazingly environment-friendly. Let's look at some numbers.

The William Gates building at Cambridge University holds computer science researchers, administrators, and a small café. Its area is 11, 110m², and its energy consumption is 2392MWh/y. That's a power per unit area of 215kWh/m²/y, or 25W/m². This building won a RIBA award in 2001 for its predicted energy consumption. "The architects have incorporated many environmentally friendly features into the building." [5dhups]

But are these buildings impressive? Next door, the Rutherford building, built in the 1970s without any fancy eco-claims——indeed without even double glazing——has a floor area of 4998m² and consumes 1557MWh per year; that's 0.85kWh/d/m², or 36W/m². So the award-winning building is just 30% better, in terms of power per unit area, than its simple 1970s cousin. Figure36-18 compares these buildings and another new building, the Law Faculty, with the Old Schools, which are ancient offices built pre 1890. For all the fanfare, the difference between the new and the old is really quite disappointing (Figure 36-18、Figure 36-19)!

Notice that the building power consumptions, per unit floor area, are in just the same units (W/m²) as the renewable powers per unit area that we discussed Comparing these consumption and production numbers helps us realize how difficult it is to power modern buildings entirely from on-site renewables. The power per unit area of biofuels is 0.5W/m²; of wind farms, 2W/m²; of solar photovoltaics, 20W/m²; only solar hot-water panels come in at the right sort of power per unit area, 53W/m².

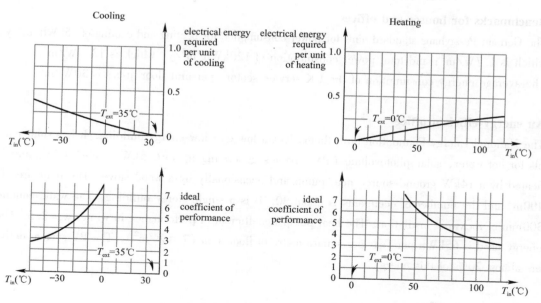

Figure 36-18 Building benchmarks.
Power used per unit area in various homes and offices.

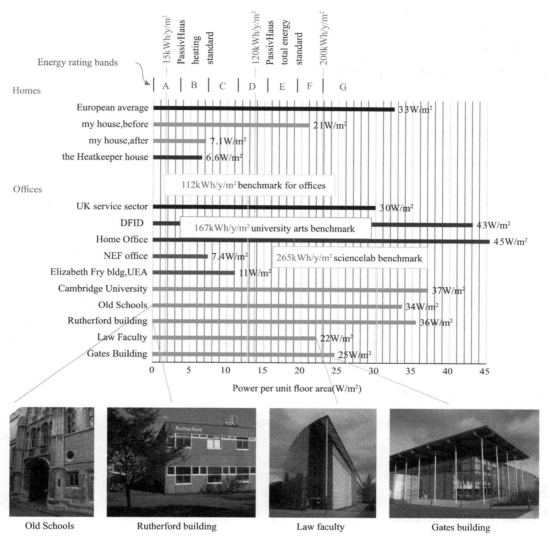

Old Schools   Rutherford building   Law faculty   Gates building

Figure 36-19 Ideal heat pump efficiencies. Top left: ideal electrical energy required, according to the limits of thermodynamics, to pump heat out of a place at temperature Tin when the heat is being pumped to a place at temperature Tout =35℃. Right: ideal electrical energy required to pump heat into a place at temperature Tin when the heat is being pumped from a place at temperature Tout = 0℃. Bottom row: the efficiency is conventionally expressed as a "coefficient of performance," which is the heat pumped per unit electrical energy. In practice, I understand that well-installed ground-source heat pumps and the best air-source heat pumps usually have a coefficient of performance of 3 or 4; however, government regulations in Japan have driven the coefficient of performance as high as 6.6.

Resource for Reference

David JC MacKay. *Sustainable Energy-without the Hot Air*. 2009: UIT Cambridge Ltd.

Section 3
Tips for Oral Presentation

Rehearsing the Presentation

There's something to be said for winging it: "Forget It!"

To present the most professional image, you need to know your presentation. It's OK to occasionally leave the main "script" but, wandering presentations that lack focus, or those too dependent on working from notes, or long pauses to compose your thoughts are never acceptable.

Rehearsing the presentation includes more than just going over what you will be saying. Rehearsing includes the entire presentation. Use the same tools too. If you are using slides, or a projector, and have access to the room you will be presenting in, rehearse there. Using a remote mouse and laser pointer for the presentation, a microphone? Rehearse the presentation with these devices.

Resource for Reference

http://www.projectorreviews.com/effectivepresentations.php

Section 4
Listening Practice

Please watch the video and answer the questions below.

1. What is the relationship between the rooms and the courtyard in Siza's house?
2. What shows the Roman feature of Siza's courtyard?
3. Why did Siza elevate the concrete block conduits that are used for supplying water and electricity to the fields of houses?

Words and Phrases:

1. social housing   社会住房（指由住房协会和地方市政会提供的廉租房或廉价房）
2. encompass   *v.*   包围；围绕
3. cobble   *v.*   粗劣地制作；修（鞋）；匆匆制作；胡乱拼凑；   *n*   鹅卵石
4. configure   *v.*   使具一定形式

5. vernacular architecture　乡土建筑
6. render　*v.*　表达
7. carnation　*n.*　康乃馨
8. portico　*n.*　柱廊，（有圆柱的）门廊
9. expropriate　*v.*　征用
10. shelve　*v.*　搁置
11. hearten　*v.*　鼓励，使振作

Resource for Reference

https://www.youtube.com/watch?v=YegwKbX06MI

Appendix | 附录

Vocabulary 词汇

A	
voracious *adj.* 贪婪的	Unit 21
accessible *adj.* 可达的	Unit 08
acknowledgement *n.* 确认、承认	Unit 05
acquaintance *n.* 熟人	Unit 05
adhesive *n.* 粘合剂；胶黏剂 *adj.* 黏着的；带黏性的	Unit 33
aesthetically *adv.* 审美地，美学上的	Unit 25
air-raid defense 人防	Unit 28
align *v.* 排成直线	Unit 16
Alpine country（France, Switzerland, and Italy） 阿尔卑斯山国家（如法国、瑞士、意大利等）	Unit 26
alternate-level 楼层变化的	Unit 17
ambiguity *n.* 含糊	Unit 06
amorphous *adj.* 形态不定的	Unit 03
amorphous *adj.* 无定形的，无组织的	Unit 18
analogous *adj.* 类似的	Unit 23
analogy *n.* 类推、类似	Unit 09
anarchy *n.* 无政府状态	Unit 09
anchor *n.* 锚；抛锚停泊；靠山；新闻节目主播 *adj.* 末棒的；最后一棒的 *vt.* 抛锚；使固定；主持节目 *vi.* 抛锚	Unit 34
anchor *v.* 锚固	Unit 08
apt *adj.* 合适的、适当的	Unit 05
archaic *adj.* 古代的；陈旧的；古体的；古色古香的	Unit 36
archetypal *adj.* 原型的	Unit 05
archetype *n.* 原型，原始模型	Unit 10
architectural dialogue 建筑对话	Unit 19
armadillo *n.* 犰狳	Unit 21
armor *n.* 盔甲，装甲	Unit 21
articulate *v.* 说清楚、接合	Unit 06
ascending magnitude 递增	Unit 04
asphalt *n.* 沥青；柏油 *vt.* 以沥青铺 *adj.* 用柏油铺成的	Unit 35
asphalt *n.* 沥青、柏油	Unit 07
astounding *adj.* 令人震惊的；令人惊骇的	Unit 35

asymmetrically　*adv.*　不均匀地，不对称地	Unit 18
at large　普遍地	Unit 02
at large　普遍地	Unit 19
Athens Charter　雅典宪章	Unit 08
atrium　*n.*　中庭	Unit 13
attribute　*n.*　属性，品质，特征	Unit 32
austerity　*n.*　紧缩；朴素；苦行；严厉	Unit 35
autonomous　自治的 独立的	Unit 35
auxiliary mean　辅助设施	Unit 14
axiom　*n.*　公理	Unit 04
B	
backbone　*n.*　支柱、脊椎	Unit 03
back-to-back　背靠背的	Unit 11
barrel　*n.*　桶	Unit 23
barrier-free design　无障碍设计	Unit 28
basement　*n.*　地下室；地窖	Unit 35
basement　*n.*　地下室	Unit 29
bay　*n.*　开间	Unit 15
be bound to　一定要	Unit 06
be enamored of　迷恋于……	Unit 24
be fused into　（被）融入……	Unit 19
be mired in　陷入	Unit 19
be permeated with　充满	Unit 21
be thick with　充满	Unit 05
bend　*v.*　弯曲	Unit 20
bicycle rack　自行车托架	Unit 33
bitumen　*n.*　沥青	Unit 34
bizarre　*adj.*　奇异的	Unit 18
blinding layer　草鞋底；地基垫层	Unit 33
bolt　*n.*　螺栓/　*v.*　拴接	Unit 20
break with　断交，决裂	Unit 01
breakthrough　*n.*　突破	Unit 27
brigade　*n.*　［军］旅	Unit 33
buckle　*v.*　弯曲，变形，起皱	Unit 24
building code　建筑规范	Unit 28
building codes　建筑规范	Unit 11
building component　建筑构件	Unit 19
built environment　建成环境	Unit 01
burnt clay　黏土砖	Unit 20

Appendix 附录 | 291

C

candlestick *n.* 烛台	Unit 06
canopy *n.* 天篷；华盖；遮篷；苍穹 *vt.* 用天篷遮盖；遮盖	Unit 36
canopy *n.* （树）冠，天篷，遮篷	Unit 30
cantilever *n.* 悬臂	Unit 18
cantilever *n.* 悬臂梁/*v.* 悬挑	Unit 21
capacity *n.* 容量	Unit 30
carnation *n.* 康乃馨	Unit 36
Carnegie Hal （美）卡耐基音乐厅	Unit 08
carport *n.* 车库；（美）车棚	Unit 36
carrying capacity 承载力	Unit 20
cast-in-place *adj.* 现浇的	Unit 27
catchment *n.* 集水处、流域	Unit 07
categorize *v.* 分类	Unit 06
category *n.* 类别	Unit 09
category *n.* 类别	Unit 22
cement *n.* 水泥	Unit 21
chaos *n.* /chaotic *adj.* 混乱/混乱的	Unit 09
charismatic *adj.* 有魅力的	Unit 19
circulation *n.* 流线	Unit 10
clad *vt.* 在金属外覆以另一种金属	Unit 31
clarity *n.* 清晰	Unit 04
cluster *n.* 簇、（住宅）组团	Unit 05
coated *adj.* 涂上一层的	Unit 29
cobble *v.* 粗劣地制作；修（鞋）；匆匆制作；胡乱拼凑； *n* 鹅卵石	Unit 36
code *n.* 法规，准则	Unit 25
coincide *v.* 一致	Unit 06
collective *adj.* 集合的，集体的	Unit 12
combination *n.* 组合，结合	Unit 25
commission *v.* 委任	Unit 27
commitment *n.* 承担义务	Unit 19
compact *adj.* 紧凑的，紧密的，简洁的	Unit 12
compartmentalization *n.* 划分、区分	Unit 09
compassion *n.* 怜悯，同情	Unit 09
complex *n.* 复合体；综合设施 *adj.* 复杂的；合成的	Unit 33
complexity/simplicity 简单/复杂	Unit 04
compressive *adj.* 压缩的	Unit 20
conceive *v.* 构想，设想	Unit 25

concourse　*n.*　汇合，集合，广场，（车站、机场）中央大厅	Unit 29
configure　*v.*　使具一定形式	Unit 36
conform to　符合，遵照	Unit 28
conical　*adj.*　圆锥的，圆锥形的	Unit 20
consequence　*n.*　结果	Unit 08
conservatism　*n.*　保守主义	Unit 02
constitute　*v.*　组成	Unit 20
contemporary　*adj.*　当代的，现代的	Unit 09
contractor　*n.*　承包人；立契约者	Unit 33
cooling tower　冷却塔	Unit 24
core　*n.*　核，核心	Unit 11
correspond to　相应、符合	Unit 05
corresponding　*adj.*　相应的	Unit 24
corridor-every-floor　每层设置走廊的	Unit 12
corrugate　*v.*　（使某物）起皱褶	Unit 09
cottage　*n.*　村舍	Unit 03
courtyard　*n.*　庭院，院子	Unit 12
crease　*n./v.*　褶皱	Unit 23
cripple　*v.*　削弱	Unit 04
criteria　*n.*　标准	Unit 10
crossover　*n.*　天桥	Unit 07
crystallize　*v.*　明确	Unit 07
curtain wall　幕墙	Unit 27
curvature　*n.*　弯曲	Unit 20
curvilinear　*adj.*　曲线的	Unit 19
cut across　抄近路	Unit 08
cylinderlike　*adj.*　柱状的	Unit 22
cylindrical　*adj.*　圆柱的，圆柱形的	Unit 20
D	
dead load　静荷载/snow load　动荷载	Unit 24
decipher　*v.*　解码，破解/decode	Unit 01
deconcentration　*n.*　分散	Unit 33
deflect　*v.*　（使）偏斜，（使）偏转	Unit 20
deliberately　*adv.*　故意地	Unit 08
deliverance　*n.*　释放，意见，判决	Unit 18
demonstrate　*v.*　证明	Unit 04
describe　*v.*　描述，画	Unit 24
detached　*adj.*　分离的，分开的；超然的　*v.*　分离	Unit 35

detached house 独立式住宅	Unit 16
developable surface 可展曲面/non-developable surface 不可展曲面	Unit 20
diagonally adv. 对角地	Unit 24
diagram n. 图解、图表	Unit 04
dime slot 投币口/dime 一角硬币	Unit 03
disabled n. 残障者	Unit 28
discipline n. 纪律	Unit 05
discretely adv. 分离的	Unit 01
displacement 位移	Unit 20
dissociation n. 分裂	Unit 09
distinguish from 区别	Unit 03
distribute v. 分布，分发	Unit 20
ditty n. 小曲、小调	Unit 04
divergent adj. 相异的，分歧的；散开的	Unit 36
domelike adj. 穹顶状的	Unit 22
double facade 双层表皮	Unit 27
double-height 两层高的	Unit 17
double-loaded adj. 内廊式的	Unit 10
Double-Orientation Unit 90° 转角单元	Unit 11
Double-Orientation Unit, open-ended 双向开敞的单元	Unit 11
downstand beam 肋形楼板梁	Unit 33
drainage mat 排水砂垫层	Unit 33
drainage n. 排水	Unit 29
drill n. 钻头	Unit 09
drop-off 下客	Unit 30
dumbbell n. 哑铃	Unit 14
dynamic adj. 动态的	Unit 03
E	
ecological adj. 生态学的	Unit 18
economy n. 简洁	Unit 04
elevation n. 高地；海拔；提高；崇高；正面图	Unit 33
emeritus adj. 名誉退休的	Unit 19
encampment n. 营地	Unit 05
enclose v. 围护/enclosure	Unit 20
enclose vt. 围绕，围合	Unit 12
encompass v. 包围；围绕	Unit 36
encompass v. 包含、环绕	Unit 08
endows with 赋予	Unit 04

endwall *n.* 端墙	Unit 23
enormity *n.* 巨大	Unit 06
enterprise *n.* 企（事）业单位；事业	Unit 09
enterprise *n.* 企业	Unit 08
entity *n.* 实体	Unit 04
envision *vt.* 想象，预想	Unit 31
esprit de corps *n.* 团体精神	Unit 27
estimated cost 估算	Unit 29
evacuation *n.* 疏散，撤退	Unit 28
evolutionary *adj.* 进化的	Unit 17
evolve *v.* 发展，进展，进化	Unit 17
exhaust air 废气；排气	Unit 34
explicit *adj.* /implicit *adj.* 外在的、清楚的/内在的、含蓄的	Unit 09
expressway *n.* 快速路	Unit 30
expropriate *v.* 征用	Unit 36
extravagant *adj.* 奢侈的，浪费的，过分的	Unit 17
F	
facade *n.* 正面；表面；外观	Unit 33
fair face 漂亮的容貌	Unit 33
fan-shaped *adj.* 扇形的	Unit 12
Fascia 筋膜	Unit 34
fascination *n.* 迷恋	Unit 09
feasible *adj.* 可行的，切实可行的	Unit 15
feature *vt.* 是…的特色，特写，放映， *vi.* 起重要作用	Unit 12
Ferrocemento *n.* 加筋水泥	Unit 21
fertile *adj.* 肥沃的, 丰富的/sterile 贫瘠的	Unit 26
finishing *adj.* 最后的，完工的， *n.* 面层	Unit 29
flexibility *n.* 弹性，适应性	Unit 18
flexible *adj.* 柔韧的，有弹性的	Unit 21
flimsy *adj.* 脆弱的	Unit 23
flush with 由于…而脸红；因…而兴奋	Unit 36
focal *adj.* 焦点的，在焦点上的；灶的，病灶的	Unit 35
folded plate 折板	Unit 23
forecourt *n.* 前院；前场	Unit 33
formwork *n.* 量规，模架；样板	Unit 33
formwork *n.* 支模材料	Unit 23
foundation *n.* 基础, 地基	Unit 29
frame *n.* 结构框架	Unit 20
framework *n.* 框架	Unit 24

Appendix 附录 | 295

G	
gallery access *adj.* 走廊进入式的	Unit 10
generator *n.* 发生器	Unit 09
geometric *adj.* 几何的，几何学的	Unit 19
glazed *adj.* 光滑的；像玻璃的；上过釉的； *v.* 上釉（glaze 的过去分词）；装以玻璃	Unit 33
glorious *adj.* 壮丽的，光辉灿烂的	Unit 25
gluttonously *adv.* 贪婪地、贪吃地	Unit 08
go to the dogs 堕落、潦倒	Unit 02
Gothic cathedral 哥特式天主教堂	Unit 25
grade *n.* 等级，坡度，斜坡，地表面与建筑物基础相遇的高度	Unit 29
gravity *n.* 重力	Unit 26
groined vault 交叉拱	Unit 25
grout *n.* 水泥浆	Unit 23
guesthouse *n.* 宾馆	Unit 18
gutter *n.* 檐沟，排水沟	Unit 22
gymnasium *n.* 体育馆/gym（缩写）/gymnastics 体操	Unit 27
gypsum board 石膏板	Unit 33
H	
hail *v.* 打招呼	Unit 07
hallway *n.* 门厅，回廊，走廊	Unit 14
hearten *v.* 鼓励，使振作	Unit 36
heat gain 热增量	Unit 27
hectares *n.* 公顷（hectare 的复数）	Unit 33
hedge *n.* 树篱，障碍物	Unit 12
heuristic *adj.* 启发式的	Unit 10
hierarchy *n.* 层级；等级制度	Unit 36
hierarchy *n.* 层级、层次	Unit 08
high-rise *adj.* 高层的	Unit 10
high-sounding *adj.* 夸张的	Unit 24
hillside housing 山地住宅	Unit 16
hollow tile 空心砖	Unit 20
homogeneity *n.* 同质；同种；同次性	Unit 35
hook on to 追随	Unit 16
hospital-cum-medical school *n.* 医院附属医学院	Unit 07
hyperbolic paraboloid 双曲抛物面/hypar（缩写）	Unit 24

I	
iconoclast *n.* 反偶像者，提倡打破旧习的人	Unit 01
illuminate *v.* 阐明、照亮	Unit 08
illustrate *v.* 用插图说明	Unit 04
impend *v.* 即将发生	Unit 09
implication *n.* 含义、暗示	Unit 06
implicitly *adv.* 含蓄地、暗中地	Unit 07
implying *vt.* 暗示，意味	Unit 12
in situ 在原地，就地；在原来位置	Unit 33
in the shape of 以……的形式	Unit 08
inclined *adj.* 倾斜的	Unit 23
incorporate into 结合/incorporate with	Unit 01
inequity *n.* 不公平、不公正	Unit 08
ingredient *n.* 成分、要素	Unit 02
ingress and egress 出入口	Unit 30
inhabitant *n.* 居民	Unit 07
inherent *adj.* 固有的，内在的	Unit 19
inspiration *n.* 灵感	Unit 18
intact 未触动的，未碰过的；未变质的	Unit 35
integral *adj.* 完整的，整体的	Unit 19
interaction *n.* 相互作用	Unit 05
interface *n.* 界面，分界面，接触面	Unit 30
interlock *vi.* 结合，连结，互锁 *vt.* 使联锁，使连结	Unit 11
interlocking *adj.* 联锁的	Unit 19
intermediate *n.* 媒介，中介	Unit 20
intersection *n.* 交叉	Unit 25
interstice *n.* 裂缝	Unit 07
interval *n.* 间隔	Unit 23
intervening *v.* 介入（intervene 的 ing 形式） *adj.* 介于中间的	Unit 35
intolerance *n.* 不能容忍	Unit 09
introverted *adj.* 内向的 *v.* 使...内向（introvert 的过去分词形式）；使内翻	Unit 35
intuitive *adj.* 直觉的	Unit 10
intuitively *adv.* 直觉地	Unit 08
inventory *n.* 详细目录	Unit 09

ironic *adj.* 讽刺的	Unit 26
irrelevant *adj.* 不相关的	Unit 05
isometric *adj.* 等距的；等角的；等轴的	Unit 33
J	
jumbotron *n.* （电视机的）超大屏幕	Unit 29
justify *v.* 证明……是适当的	Unit 24
K	
ketch *n.* 双桅帆船	Unit 21
knight *n.* 骑士	Unit 21
L	
lacquer *n.* 漆；漆器 *vt.* 涂漆；使表面光泽	Unit 35
land premium 土地价格	Unit 29
lattice *n.* 网格	Unit 02
lean concrete 贫混凝土，少灰混凝土	Unit 34
lest *conj.* 以免	Unit 06
limp *v.* 跛行	Unit 07
linoleum *n.* 油布；油毯；漆布	Unit 33
literally *adv.* 差不多	Unit 07
load *n.* 荷载	Unit 20
longitudinal *adj.* 经度的，纵向的	Unit 14
longitudinal *adj.* 纵向的	Unit 23
low-rise *adj.* 低层的	Unit 10
luminaires *n.* 灯具（luminaire 复数）；照明器	Unit 33
M	
magnificent *adj.* 宏伟的	Unit 24
maisonette *n.* 小房屋，出租房间	Unit 15
make a distinction 区别、区分	Unit 02
mandatory *adj.* 强制性的	Unit 14
mania *n.* 狂热	Unit 08
manifesto *n.* 宣言	Unit 18
mantelpiece *n.* 壁炉架	Unit 06
marvel *n.* 奇迹	Unit 21
masonry *n.* 石工；石工行业；石造建筑	Unit 34
matrix *n.* 矩阵	Unit 12
matte *adj.* 不光滑的，表面粗糙的	Unit 12
medieval *adj.* 中世纪的	Unit 21
megastructure *n.* 巨型结构	Unit 01
melancholy *adj.* 忧郁的	Unit 27

melt into 融入	Unit 07
memorial *adj.* 纪念的，悼念的/*n.* 纪念碑，纪念物	Unit 09
mental construct 心智构造	Unit 09
metaphor *n.* 隐喻	Unit 01
metropolis *n.* 大都市	Unit 03
metropolis *n.* 都会，大城市	Unit 28
Metropolitan Opera House （美）大都会歌剧院	Unit 08
militant *adj.* 激进的	Unit 09
mill *v.* 转悠	Unit 03
mineral-fibre 矿物纤维	Unit 33
molecule *n.* 分子	Unit 03
mollusk *n.* 软体动物	Unit 21
monolithic *adj.* 整体的	Unit 20
monotony *n.* 单调、千篇一律	Unit 03
mortar *n.* 灰浆	Unit 21
multiple-access 多个出入口的	Unit 16
multiplicity *n.* 多样性	Unit 06
municipality *n.* 市政当局	Unit 08
N	
narrative *n.* 叙述	Unit 19
natatorium *n.* 游泳场，游泳池	Unit 28
natural city 自然城市/artificial city 人造城市	Unit 02
neuroscience *n.* 神经系统科学	Unit 18
neutral *adj.* 中性的，中立的	Unit 23
newsrack *n.* 报栏	Unit 03
nibs *n.* 大人物；上司；傲慢的人	Unit 34
nodal *adj.* 节（点）的	Unit 06
non-loadbearing 非承重的	Unit 33
northerly *adj.* 北方的，向北的；来自北方的 *n.* 北风 *adv.* 向北；来自北方	Unit 33
O	
ominous *adj.* 不吉利的、凶兆的	Unit 09
onset *n.* 肇始	Unit 02
open-ended *adj.* 末端开口的	Unit 14
optimization *n.* 最佳化，最优化	Unit 32
optimum *adj.* 最适宜的	Unit 14
outrage *n.* 暴行、侮辱	Unit 02
overlap *v.* 交搭，叠盖	Unit 15

overlap *v.* 重叠，交迭/superimpose	Unit 01
overwhelming *adj.* 压倒性的	Unit 08
P	
palatial *adj.* 富丽堂皇的	Unit 14
paradigm *n.* 范例	Unit 10
paradigm *n.* 范例，范型	Unit 01
parallelogram *n.* 平行四边形	Unit 09
parapet *n.* 栏杆；扶手；矮护墙	Unit 34
paraphernalia *n.* 随身用具	Unit 31
parti *n.* 构图	Unit 13
party wall *n.* 分户墙	Unit 13
patina *n.* 光泽	Unit 02
pedestrian circulation 步行流线	Unit 07
perforate *vt.* 穿孔于	Unit 31
perforated *adj.* 穿孔的；有排孔的 *v.* 穿孔（perforate 的过去分词）	Unit 33
performing art 表演艺术	Unit 08
periodic *adj.* 周期的	Unit 16
peripheral *adj.* 外围的	Unit 13
permutation *n.* 改变，交换	Unit 13
perpendicular *adj.* 垂直的	Unit 11
perpendicular *adj.* 垂直的/vertical *adj.* 竖直的	Unit 20
pertinent *adj.* 有关的，中肯的	Unit 18
picturesque *adj.* 独特的、风景如画的	Unit 03
pigeonhole *n.* 鸽巢	Unit 09
pilaster 壁柱，半露柱	Unit 33
pile foundation 打桩基础	Unit 33
piling *n.* 打桩，打桩工程	Unit 29
Pinwheel *n.* 风车	Unit 13
pipes and ducts 管线	Unit 20
pivot *n.* 枢轴；中心点；旋转运动 *adj.* 关键的 *vt.* 以…为中心旋转 *vi.* 在枢轴上转动；随…转移	Unit 36
plastic *adj.* 造型的	Unit 03
plastic *adj.* 造型的	Unit 09
plate *n.* 板	Unit 19
plausible *adj.* 似是而非的	Unit 10
plinth *n.* 柱基；底座	Unit 33
plumbing *n.* 管道工程	Unit 11
plumbing *n.* 管道工程	Unit 29

pneumatic *adj.* 充气的	Unit 27
popularity *n.* 普及，流行	Unit 27
popularized *adj.* 普及的，大众化的	Unit 18
portico *n.* 柱廊，（有圆柱的）门廊	Unit 36
precast *adj.* 预制的 *vt.* 预制；预浇制	Unit 33
predisposition *n.* 素质、倾向	Unit 08
prefabricated *adj.* 预制构件的 *v.* 预先建造组合（prefabricate 的过去分词）	Unit 33
prefabricated *adj.* 预制的	Unit 27
preponderantly *vi.* 占优势，超过，胜过	Unit 18
prescription *n.* 规定	Unit 25
prestressed concrete 预应力混凝土	Unit 27
primitive *adj.* 简单的、原始的	Unit 09
proliferation *n.* 增殖，大量出现	Unit 14
prompt *v.* 促使	Unit 08
proportions *n.* 比例；大小（proportion 的复数形式）	Unit 36
proprietary *adj.* 所有的，私人拥有的	Unit 32
prototypes *n.* 原型	Unit 10
provoke *v.* 煽动；激起	Unit 09
prowl *v.* 巡游、徘徊	Unit 07
psychological *adj.* 心理（上）的/spiritual 精神上的	Unit 01
pub-crawl *v.* 逛酒店喝酒	Unit 07
public address and communication system 公共广播系统	Unit 29
pulled thread 拉索	Unit 24
putty *n.* 油灰；氧化锡 *vt.* 用油灰填塞	Unit 34
R	
racially *adv.* 按人种	Unit 18
radius *n.* 半径/diameter 直径	Unit 26
rammed earth 夯土	Unit 01
rapid-setting mortar 快干砂浆	Unit 27
rear *n.* 后面，后方 *adj.* 后面的	Unit 11
receptacle *n.* 容器	Unit 03
recess *n.* 休息；休会；凹处 *vt.* 使凹进；把…放在隐蔽处 *vi.* 休息；休假	Unit 34
recess *n.* （墙壁等的）凹进处， *vt.* 使凹进	Unit 30
recreation *n.* 娱乐	Unit 07
recreational *adj.* 娱乐的，消遣的；休养的	Unit 33
rectangular *adj.* 矩形的	Unit 25
redevelopment *n.* 再开发、	Unit 07
regrettable *adj.* 可惜的	Unit 25
reinforced concrete 钢筋混凝土	Unit 20
reluctance *n.* 勉强、不愿意	Unit 02
remedy *n.* 疗法、解决问题的手段	Unit 02
render *v.* 表达	Unit 36

rendered　*v.*　提出；描绘（render 的过去分词）；放弃；报答；　*adj.*　已渲染的	Unit 33
renewal　*n.*　更新/renovation/revitalization	Unit 01
renowned　*adj.*　著名的，有声誉的	Unit 19
residential　*n.*　住宅的	Unit 27
residue　*n.*　残留物	Unit 05
resistant　*adj.*　抵抗的，反抗的，耐久的	Unit 19
retaining wall　挡土墙	Unit 13
revenue　*n.*　税收	Unit 08
rhyme　*n.*　韵律、押韵的诗词	Unit 04
rhythmic　*adj.*　有韵律的，有节奏的；格调优美的	Unit 27
rift　*n.*　裂缝	Unit 09
rigidity　*n.*　严格	Unit 05
ring road　环路	Unit 30
rock outcroppings　岩石露头	Unit 18
roofing material　屋面材料	Unit 20
rostrum　*n.*　讲坛，演讲坛	Unit 28
rotational hyperboloid　旋转双曲面	Unit 24
rowhouse　*adj.*　联排式住宅	Unit 10
S	
saddlelike　*adj.*　马鞍状的	Unit 22
sanitation　*n.*　卫生，卫生设施	Unit 28
scaffold　*n.*　脚手架	Unit 20
scheme　*n.*　配置，计划	Unit 11
schizophrenia　*n.*　精神分裂症	Unit 09
screed　*n.*　冗长的文章；[地质]砂浆层	Unit 33
section　*n.*　截面	Unit 33
seismic design　抗震设计	Unit 28
self-contained　*adj.*　设备齐全的，独立的	Unit 15
self-proclaimed　*adj.*　自称的	Unit 01
semiprivate　*adj.*　半私密的	Unit 16
Separating layer　分离层，分层，分隔层	Unit 33
shear　*v.*　剪切	Unit 20
shed　*n.*　棚，库	Unit 27
shelve　*v.*　搁置	Unit 36
shield　*n.*　盾	Unit 27
shutters　*n.*　百叶门窗，百叶窗（shutter 的复数形式）	Unit 36
signage　*n.*　标识	Unit 32
signify　*v.*　表示，意味	Unit 19
sills　*n.*　[建]基石（sill 的复数）	Unit 34
silver lining　（不幸或失望中的）一线希望	Unit 24
simpleminded　*adj.*　头脑简单的	Unit 08

simultaneously　*adv.*　同时地	Unit 09
single out　挑选（出）	Unit 01
single-loaded　*adj.*　外廊式的	Unit 10
Single-orientation Unit　单一朝向的单元	Unit 11
single-run　*n.*　单跑楼梯	Unit 15
site plan　总设计图；总平面图；修建性详细规划	Unit 33
siting　*n.*　选址	Unit 10
skew　*adj.*　歪斜的	Unit 20
skip-stop corridor　隔层设置的走廊	Unit 10
slab　*n.*　板式建筑	Unit 10
slab　*n.*　厚板	Unit 29
slate　*v.*　用石板瓦盖，用板岩覆盖（例如屋顶）；指定	Unit 18
slope　*n.*　坡度	Unit 27
slot　*n.*　缝，槽	Unit 14
social housing　社会住房（指由住房协会和地方市政会提供的廉租房或廉价房）	Unit 36
solitary　*n.*　独居者；隐士　*adj.*　孤独的；独居的	Unit 35
spacer　*n.*　垫片；[遗]间隔区；逆电流器	Unit 34
span　*n.*　跨度/*v.*　跨越	Unit 20
spandrel　*n.*　拱肩	Unit 35
spatial sequence　空间序列	Unit 02
spherical　*adj.*　球的，球形的	Unit 20
spine　*n.*　脊柱	Unit 22
spiral staircase　螺旋楼梯	Unit 27
split-level　*adj.*　房内有不同高度平面的，错层的	Unit 13
spontaneously　*adv.*　自然地、自发地	Unit 02
spout　*n.*　喷口；水龙卷；水落管；水柱　*vt.*　喷出；喷射	Unit 34
spraygun　*n.*　喷枪/airbrush 喷笔	Unit 27
square/piazza　*n.*　广场/plaza	Unit 03
square-rigger sail　*n.*　横帆船	Unit 26
stack　*n.*　堆，　*v.*　堆叠	Unit 11
stage　*vt.*　上演，举行	Unit 28
staggered-plan　交错变化的平面	Unit 13
stand　*n.*　看台	Unit 31
stem　*v.*　生长	Unit 06
sterility　*n.*　贫乏	Unit 01
stiff　*adj.*　有刚度的/strong　*adj.*　有强度的	Unit 20
stimulate　*v.*　刺激；激励	Unit 09
stock-in-trade　*n.*　存货，惯用手段	Unit 18
strand　*n.*　线、索	Unit 09
strategic　*adj.*　战略的，战略上的	Unit 32
strategy　*n.*　策略	Unit 12

strut　*n.*　压杆	Unit 24
subjective　*adj.*　主观的/objective　客观的	Unit 19
subservient　*adj.*　屈从的；奉承的；有用的；有帮助的	Unit 36
substantially　*adv.*　充分地，实质上地	Unit 20
substrate　*n.*　基质；基片；底层（等于 substratum）；酶作用物	Unit 33
subtle　*adj.*　微妙的	Unit 04
succinctly　*adv.*　简洁地	Unit 04
suites　*n.*　套房	Unit 29
superfluous　*adj.*　多余的	Unit 20
sustenance　*n.*　生存、生计、维持	Unit 08
symmetrical　*adj.*　对称的	Unit 06
symmetrical　*adj.*　对称的	Unit 22
symphony　*n.*　交响乐	Unit 06
Synopsis　*n.*　大纲	Unit 28
T	
take up　占据	Unit 11
tectonic　*adj.*　构造的；建筑的；地壳构造上的	Unit 36
tensile　*adj.*　可拉长的，拉力的	Unit 21
tension　*n.*　张力，拉力	Unit 23
terrace houses　退台式住宅	Unit 16
terrain　*n.*　［地理］地形，地势；领域；地带	Unit 35
tesseract　*n.*　立方体的四维模拟，超正方体	Unit 18
theatrical　*adj.*　夸张的，戏剧性的；	Unit 09
thematic　*adj.*　主题的	Unit 32
thin shell　薄壳	Unit 25
thrust　*n.*　推力	Unit 23
tie-rod　*n.*　系杆	Unit 23
tile　*n.*　面砖	Unit 20
tilt　*v.*　（使）倾斜	Unit 24
toddle　*v.*　东倒西歪地走路、蹒跚学步	Unit 07
tower　*n.*　塔式建筑	Unit 10
transfer structures　转换层结构	Unit 33
transverse　*adj.*　横的	Unit 11
transverse　*adj.*　横断的	Unit 23
truss　*n.*　构架，桁架，钢梁	Unit 20
truss　*n.*　桁架	Unit 31
turnover　*n.*　营业额	Unit 32
turnstile　*n.*　十字转门，转栅	Unit 30
twisting mechanism　扭转机制	Unit 20
twofold　*adj.*　双重的	Unit 23
typological　*adj.*　类型（学）的	Unit 10
typology　*n.*　类型学	Unit 10

U	
undulating　*adj.*　起伏的	Unit 26
Unique　*adj.*　独一无二的	Unit 08
upside-down　*adj.*　颠倒的	Unit 20
V	
valiant　*adj.*　勇敢的	Unit 02
vapour barrier　隔汽层	Unit 33
vault　*n.*　拱顶	Unit 19
vehicle for thought　*n.*　传达思维的手段	Unit 09
Venn diagram　文氏图，用于显示元素集合重叠区域的图示	Unit 18
ventilation　*n.*　通风	Unit 11
ventilation　*n.*　通风	Unit 29
véranda　游廊，走廊，凉台间	Unit 35
vernacular　*adj.*　本土的，本国的，地方的	Unit 01
vernacular architecture　乡土建筑	Unit 36
vestibule　*n.*　门廊，前厅	Unit 30
veto　*v.*　投票反对/vote　投票赞成	Unit 25
virtually　*adv.*　事实上、实质上	Unit 05
vista　*n.*　狭长的景色，街景	Unit 31
visualize　*v.*　想象，形象化	Unit 09
visualize　*v.*　形象化	Unit 23
vital　*adj.*　至关重要的	Unit 08
voracious　*adj.*　贪婪的	Unit 21
W	
walk-up　*n.*　无电梯的公寓，*adj.*　无电梯的	Unit 10
walled　*adj.*　有墙壁的	Unit 12
watertight　*adj.*　水密的/waterproof　防水的	Unit 26
weld　*n.*　焊缝/　*v.*　焊接	Unit 20
welded mesh　焊接铁丝网	Unit 21
wino　*n.*　酒鬼	Unit 09
wreck　*v.*　破坏、拆毁	Unit 02
Y	
yardstick　*n.*　码尺	Unit 22
yearn for　渴望	Unit 03
Z	
Zen　禅（佛教）	Unit 01
zone　*n.*　分区	Unit 01
zoning　*n.*　分区	Unit 08

Notes for Video Material 音像资料注解

Artifacts: Sacred Spaces

Episode I

1. final touch   最后的润色
2. masterpiece   杰作
3. temple of Heaven   天坛
4. swooping   倾斜的
5. stately   庄严的
6. cosmological symbolism   宇宙哲学象征主义
7. tyrant   暴君
8. tombmound   坟冢
9. tactician   战术家
10. annihilate   消灭
11. grandeur   宏伟
12. pharaoh   法老
13. mercury   水银
14. terracotta   陶瓦 terracotta soldier/pottery/ceramic/porcelain
15. provide for   供奉
16. deceased   死者/disease
17. granary   谷仓

Episode II

1. chrome   铬（克罗米）
2. stonehenge   史前巨石柱
3. joinery   细木工
4. flood plain   冲积平原
5. piling   打桩工程/pile driver   打桩机
6. platform   台基
7. stone base   石础
8. podium   墩墙、基座
9. bay   间
10. supporting/enclosing
11. plague   折磨
12. upheaval   巨变、隆起
13. spire   尖顶
14. facade   立面/elevation

Episode III

1. commission   委托
2. missionary   传教士
3. patronize   资助
4. imperial   皇帝的
5. ornament   装饰
6. denote   表示
7. ridge-devouring breast   饕吻兽
8. elements   自然环境
9. glazed tile   琉璃瓦
10. pole/column/pillar   柱子
11. lacquer   漆、漆器
12. bracket   托架
13. overhang   悬挑、出檐
14. see-saw   秋千，跷跷板
15. fulcrum   支点

Episode IV

1. flying buttress   飞扶壁
2. myriad   无数的
3. methods and forms of architecture   营造法式
4. treatise   论文
5. Shaman   道士、巫师
6. dictate   规定、指示
7. site   选址
8. steep   浸泡
9. animistic   万物有灵论的
10. lodestone   天然磁石
11. constellation   星座
12. dipper   北斗七星
13. diviner   巫师
14. red phoenix/white tiger/green sea dragon/black tortoise   朱雀/白虎/青龙/玄武

Episode V

1. acupuncture   针灸
2. auspicious   幸运的
3. wing chime   风铃
4. meandering   蜿蜒的
5. immortal   仙人的，不朽的

Episode VI

1. courtyard house   四合院

2. heavenly well 天井
3. fabulous 神话般的，传奇的
4. cosmopolitan 大都市
5. pagoda 佛塔
6. Stone rubbing 拓本
7. mural 壁画
8. gridiron 格状物
9. curfew 宵禁

Episode VII
1. thoroughfare 大道
2. altar 祭坛
3. ancestral temple 宗祠
4. mosque 清真寺
5. cater to 投合
6. Buddha 菩萨，佛
7. Sanskrit 梵语，梵语的
8. Sutras 佛经
9. Stupa 浮屠
10. Zen Buddhism 禅宗

Episode VIII
1. abbot of the order. （寺庙）主持
2. legacy 遗产
3. tablet 碑，牌匾
4. nestle 安置
5. quarry 采石场
6. headstone 墓石
7. wrest 夺取
8. heir 继承人
9. filial 子女的，孝顺的
10. piety 虔诚，孝行

Episode IX
1. Khubilai 忽必烈
2. Khan 可汗
3. knob 门球，突出物
4. pole star 北极星
5. celestial 天上的
6. pivotal 枢纽的，关键的
7. terrestrial 人间的
8. mantle 斗篷

9. mandate 委任，训令
10. solstice 至日
11. homage 敬意
12. intermediary 中介
13. acoustics 声学效果
14. Temple of annual pray/Temple of Universe/Round Mound 祈年殿/皇穹宇/圜丘
15. don 穿上
16. prostrate 下跪
17. impart 传达、告知

Skyscrapers：going up

Episode I
1. behind the scene 幕后
2. prowess 威力/power
3. incorporate 结合
4. agent 代理，乙方/client 客户、甲方
5. arabesque 阿拉伯式图案
6. nestle 安置
7. plight 困境
8. indigenous 本土的/domestic

Episode II
1. pit 凹坑
2. jerky 快速拉动
3. slosh 泼溅
4. nauseous 令人恶心的
5. literally 差不多，简直
6. friction 摩擦
7. microscopic 显微的，
8. precarious 不稳定的，不安全的
9. sliding bearing 滑动支承
10. pin 钉住
11. expansion joint 弹性连接
12. gondola 冈朵拉（两头尖的小船），吊篮

Episode III
1. laminated glass 平板玻璃
2. top off 结束，收头
3. silhouette 轮廓，天际线

Episode IV
1. churn 搅拌

2. inferno 地狱
3. scrap 废料
4. furnace vessel 熔炉
5. blast oxygen furnace
6. masonry 砖石建筑，石工
7. pulley 滑轮
8. winch 绞盘

Episode V

1. ground zero （核爆炸）爆心
2. Goliath ［圣经］被牧羊人大卫杀死的巨人
3. rivet 铆钉
4. gargantuan 巨大的
5. tenant 房客
6. fabulous 难以置信的
7. observatory 观象台
8. vertigo 眩晕
9. augment 增加
10. intercom 内部对讲系统

Episode VI

1. gel 凝胶
2. regimen 养生之道
3. custodian 管理员
4. own up to 承认

Episode VII

1. pristine 质朴的
2. stark 刻板的
3. uninspired 平庸的
4. girder 大梁
5. counterweight 配重，平衡物
6. flamboyant 华丽的，绚烂的

Episode VIII

1. do away with 废除
2. lunatic 精神错乱的
3. metallurgy 冶金术
4. slot machine 自动贩卖机，吃角子老虎机

Episode IX

1. flashy 浮华的
2. double as 兼具……

Super Structures of America: Reach for the Sky-Los Angeles Concert Hall

Episode I
1. visionary *adj.* 幻影的，幻想的，梦想的
2. notion *n.* 概念，观念，想法
3. sensuous *adj.* 感觉上的，给人美感的
4. defy *vt.* 不服从，反抗

Episode II
1. cardboard *n.* 纸板
2. tin *n.* 锡，马口铁
3. obsolete *adj.* 荒废的，陈旧的
4. whiz kid *n.* 神童，优等生
5. futuristic *adj.* 未来派的

Episode III
1. prefabricate *v.* 预制
2. fluid *adj.* 流动的，不固定的，可改变的
3. principal *n.* 负责人

Episode IV
1. CAM 计算机辅助制造（Computer Aided Manufacturing）
2. flange *n.* 边缘
3. freestanding *adj.* 独立式的，不需依靠支撑物的
4. sail *n.* 帆
5. billowing *adj.* 巨浪似的，汹涌的

Super Structures of America: Reach for the Sky-Las Vegas

Episode I
1. neon *n.* 霓虹灯
2. gaudiest *adv.* 俗丽地
3. glitzy *adj.* 闪光的，耀眼的，眩目的
4. tacky *adj.* 俗气的
5. outrageous *adj.* 蛮横的，令人不可容忍的
6. visually *adv.* 在视觉上地，真实地
7. frenzy *n.* 狂暴，狂怒

Episode II
1. mask *vt.* 戴面具，掩饰，使模糊
2. fancy *adj.* 奇特的，异样的
3. lavishly *adv.* 浪费地，丰富地
4. run down 逐渐恶化
5. seedy *adj.* 破旧的，褴褛的，不适的

6. chapel   n.   小礼拜堂，礼拜
7. stuff   n.   东西

Episode III

1. grim   adj.   严酷的
2. run out of steam   泄气，失去势头
3. ruthlessly   adv.   冷酷地，残忍地
4. make way   前进，让路
5. developer   n.   开发者，开发商
6. juxtaposition   n.   毗邻，并置，并列
7. spiffer   adj.   好看的，漂亮的，出色的

Episode IV

1. stroll   n.   漫步，闲逛
2. countdown   n.   倒数计秒
3. transplant   v.   移民，迁移
4. similarity   n.   类似，类似处
5. exotic   adj.   异国情调的，外来的，奇异的
6. theme park   主题公园
7. ramp   n.   斜坡，坡道
8. ornate   adj.   装饰的，华丽的
9. Venetian   adj.   威尼斯的
10. dusk   n.   薄暮，黄昏

Episode V

1. curvature   n.   弯曲，曲率
2. scale down   按比例减少
3. steering   n.   操纵，掌舵
4. pedal   n.   踏板

Discovery Magazine I: Engineering Disasters-Manhattan's Citicorp Center

Episode I

1. landmark   n.   地标
2. identity   n.   身份，特性
3. congregation   n.   集合，集会
4. tear down   扯下，拆卸
5. air right   空权
6. stilt   n.   支柱

Episode II

1. wind tunnel   风洞
2. pseudo   adj.   假的，冒充的
3. canyon   n.   峡谷

Episode III

1. whirl　　*n.*　　旋转
2. eddy　　*n.*　　旋转，漩涡
3. swirling　　*v.*　　使成漩涡，打漩
4. crank up　　加快，做好准备
5. turbulence　　*n.*　　动荡，紊乱
6. dampen　　*v.*　　使潮湿，使沮丧
7. hydraulic　　*adj.*　　水力的，水压的

Episode IV

1. levitate　　*v.*　　（使）轻轻浮起，（使）飘浮空中
2. actuator　　*n.*　　激励者
3. magnitude　　*n.*　　大小，数量，巨大，广大，量级

Episode V

1. oddly　　*adv.*　　奇特地，古怪地
2. diagonal　　*adj.*　　斜的，斜纹的，对角线的
3. beam　　*n.*　　梁
4. stress　　*n.*　　压力
5. whatsoever　　*pron.*　　无论什么

Episode VI

1. weld　　*vt.*　　焊接
2. strain　　*n.*　　张力
3. bol　　*tv.*　　螺钉拴住
4. scary　　*adj.*　　引起惊慌的
5. shear the bolts　　剪切螺钉

Episode VII

1. bend　　*v.*　　弯曲，
2. fortress　　*n.*　　堡垒，要塞
3. shudder　　*vi.*　　战栗，发抖
4. plummet　　*n.*　　铅锤，重荷　　*vi.*　　垂直落下
5. elevator shaft　　电梯井

Episode VIII

1. vulnerable　　*adj.*　　易受攻击的
2. dilemma　　*n.*　　进退两难的局面
3. mobilize　　*v.*　　动员

Episode IX

1. celebrity　　*n.*　　名声，名人
2. pariah　　*n.*　　贱民（印度的最下阶级）
3. contemplate　　*v.*　　凝视，沉思
4. client　　*n.*　　客户

5. face up to　勇敢地面对
6. tuck　*n.*　缝摺　*vt.*　打摺　*vi.*　折成摺子
7. sanctuary　*n.*　避难所

Resources for Reference

Artifacts：Sacred Spaces
http://www.tudou.com/programs/view/ioWWJU7-p5A/

Skyscrapers：going up
http://www.hollywood.com/tv/Skyscrapers_Going_Up/5191233

Super Structures of America：Reach for the Sky
http://www.verycd.com/files/04679348466500054dec22854cc63dd5750

Discovery Magazine I：Engineering Disasters
https://www.cmule.com/viewthread.php?tid=132047&extra=page%3D6

Bibliography 参考文献

1. 孙万彪. 中级翻译教程 [M]. 上海外语教育出版社, 2003. 2.
2. 孙万彪, 王恩铭. 高级翻译教程 [M]. 上海外语教育出版社, 2000. 12.
3. 夏唐代. 建筑工程英语 [M]. 华中科技大学出版社, 2005. 4.
4. (美) 斯坦, (美) 斯普雷克尔迈耶. 建筑经典读本——国外高等院校建筑学专业教材 [M]. 王群等译. 中国水利水电出版社, 2004. 3.
5. Richard Le Gates & Frederic Stout. The City Reader, 1996: Routledge, London, pp 118-131
6. Roger Sherwood. *Modern Housing Prototypes*. 2002: Harvard University Press
7. Christopher Alexander. *A Pattern Language*. 1977: Oxford University Press
8. Adrian Forty. *Words and Buildings: A Vocabulary of Modern Architecture*. 2004: Thames & Hudson
9. Andrea Deplazes. *Constructing Architecture, Materials, Processes, Structures*. 2005: Birkhäuser.
10. David JC MacKay. *Sustainable Energy-without the Hot Air*. 2009: UIT Cambridge Ltd.
11. http://www.tudou.com/programs/view/ioWWJU7-p5A/
12. http://www.hollywood.com/tv/Skyscrapers_Going_Up/5191233
13. http://www.patternlanguage.com
14. http://www.arcosanti.org
15. http://www.uic.edu/~pbhales/Levittown.html
16. http://www.aviewoncities.com/ny
17. http://www.library.cornell.edu/Reps/DOCS/sitte.htm
18. http://eng.archinform.net
19. http://en.wikipedia.org/wiki/Pier_Luigi_Nervi#Biography
20. http://www.greatbuildings.com
21. http://en.wikipedia.org
22. http://www.structurae.de
23. http://www.eb.com
24. http://www.richardrogers.co.uk/theory/public_domain
25. http://www.richardrogers.co.uk/theory/flexibility
26. http://www.arcspace.com/architects/Steven_Holl/beijing/index.htm
27. http://www.oma.eu/index.php?option=com_projects&view=project&id=55&Itemid=10
28. http://www.calatrava.info
29. http://www.encyclopedia.com/doc/1P1-104828612.html
30. http://www.hup.harvard.edu/catalog/SHEMOD.html
31. http://www.seattle.gov/DPD/
32. http://www.amazon.com
33. http://www.patternlanguage.com
34. http://www.frieze.com/issue/article/high_life
35. http://www.nytimes.com/2008/06/08/magazine/08mvrdv-t.html?pagewanted=1&fta=y
36. http://www.mvrdv.nl/_v2/
37. http://www.oma.eu/

38. http://www.eu-greenbuilding.org/
39. http://www.greenbuilding.com/
40. http://www.nyc.gov/html/dcp/html/subcats/zoning.shtml
41. http://www.cityofchicago.org/Zoning/
42. http://geography.about.com/od/urbaneconomicgeography/a/zoning.htm
43. http://www.fpza.org/index.shtml
44. http://www.ci.cambridge.ma.us/cdd/cp/zng/zord/index.html
45. https://www.youtube.com/watch?v=dSfkim0mohA
46. https://www.youtube.com/watch?v=OmjrTghS6gw
47. https://www.youtube.com/watch?v=iUG1rtmAyxc
48. https://www.youtube.com/watch?v=YegwKbX06MI
49. Architect's Essentials of Presentation Skills, David Greusel, AIA John W: ley & Sons, Iuc, 2002, New York.
50. Writing For Design Professionals. Stephan A. Kliment WW. NoRTON & COMPANP. New York. London. 2006.